# UGLY
# WAR
## PRETTY PACKAGE

# UGLY
# WAR

## PRETTY PACKAGE

How CNN and Fox News
Made the Invasion of Iraq
High Concept

**DEBORAH L. JARAMILLO**

*Indiana University Press*
BLOOMINGTON AND INDIANAPOLIS

Publication of this book is made possible in part with the assistance of a Challenge Grant from the National Endowment for the Humanities, a federal agency that supports research, education, and public programming in the humanities.

This book is a publication of

Indiana University Press
601 North Morton Street
Bloomington, IN 47404-3797 USA

www.iupress.indiana.edu

*Telephone orders*   800-842-6796
*Fax orders*         812-855-7931
*Orders by e-mail*   iuporder@indiana.edu

∞ The paper used in this publication meets the minimum requirements of the American National Standard for Information Sciences—Permanence of Paper for Printed Library Materials, ANSI Z39.48-1992.

Manufactured in the United States of America

Library of Congress Cataloging-in-Publication Data

Jaramillo, Deborah Lynn, date-
  Ugly war, pretty package : how CNN and Fox News made the invasion of Iraq high concept / Deborah L. Jaramillo.
      p. cm.
  Includes bibliographical references and index.
  ISBN 978-0-253-35363-4 (cloth : alk. paper) — ISBN 978-0-253-22122-3 (pbk. : alk. paper)  1. Iraq War, 2003—Television and the war. 2. Television broadcasting of news—United States. 3. War in mass media. 4. Mass media—Objectivity—United States. I. Title.
  DS79.76.J373 2009
  956.7044'31—dc22
                                2009008963

1  2  3  4  5  14  13  12  11  10  09

*To my Harlingen family,*
*my Austin family, and the*
*family I fell in love with*
*in Tucson*

# Contents

# Acknowledgments

As the completion of this project took several years of my life, I have a number of wonderful people to thank. First, I wish to acknowledge the inspired guidance of Dr. Mary Kearney, who encouraged me to "embrace the crisis" and find the research topic that held meaning for me. Fortunately, two Ford Foundation fellowships funded the embrace, the research, and the writing, while Dr. Kearney and Dr. John Downing facilitated the improvement of that writing. A million thanks to three different sources of news footage: Dr. Jacqueline Sharkey at the University of Arizona, who, without having met me in person, went out of her way to provide me with tapes of Fox News; the amazing staff at the media library in the University of Texas at Austin College of Communication, who provided me with tapes of CNN; and the staff at the Vanderbilt Television News Archive, who filled in the gaps. Of course, thank you to the amazing scholars who reviewed the manuscript and offered invaluable suggestions. And thank you to everyone at Indiana University Press for their dedication to this project.

The years of creation required a great deal of personal help, as well. At the top of the list is Rodolfo Fernández, who lovingly tolerated my moods and my erratic sleep schedule in the LGA and who was an incredible sounding board for my ideas. I do not deserve him. My parents, Frank and Dora, always supported my educational pursuits and never underestimated me. Their successes in life—mutual and individual—dwarf my achievements. My brother JR is the true writer of the family, and his genius is always on my mind. Dr. Susan McLeland has been a role model for me since my undergraduate years, and I would be lost without her compassion and unpretentious intellect. Rosamar Torres, better than any sister could be, made my third and fourth years of doctoral work two of the greatest years of my life. Holly Custard and Adrian Johnson have been constant and selfless sources of support, love, wisdom, and housing. I am humbled by their profound humanity. Mark Cunningham, my angel and co-conspirator, astounds me with his writing and blesses me with his friendship. My University of Arizona mentors—Dr. Beretta Smith-Shomade, Dr. Caryl Flinn, and Dr. Eileen Meehan—will always represent the personal generosity and intellectual rigor to which I aspire. Most recently, Sky Sitney, Sarah Bilodeau, and Adrian Spencer of SILVERDOCS taught me how to step outside of academia and work like mad for the pure love of cinema. Finally, I have to acknowledge every single image that appears on my beautiful LCD screen. Enlightening or embarrassing, entertaining or infuriating, TV never stops inspiring me to write.

# INTRODUCTION

## *The Spectacle of Televised War*

War is a fabrication and the media elevate it to the level of spectacle. So goes the theme of Barry Levinson's film *Wag the Dog* (1997), a Hollywood production about a Hollywood production. In the film, the president of the United States is implicated in a sex scandal days before a presidential election. In order to deflect attention from the president's alleged affair with an underage girl, his staff hires a shady political player to create a diversion. His solution is to persuade a Hollywood producer to stage a war. The result is replete with almost every motif that emerged in the 1991 Persian Gulf War, which, according to Robert De Niro's character, was also a fake. The film is a brilliantly cynical send-up of patriotism and politicians and the ways the media wrap themselves up in both. The plot of *Wag the Dog* is a cinematic version of Jean Baudrillard's (1995) argument that the televisuality of the 1991 war coverage eclipsed the actual events. I refer to Levinson's film not to argue that Hollywood masterminded the 1991 war or the 2003 invasion at the behest of presidential administrations. Instead, I argue that institutions and organizations can and do represent, negotiate, and explain war in Hollywood terms.

*Wag the Dog* posits that a formula for mediating war exists. Gripping images of young innocents, patriotic songs, imperiled heroes, and nationalistic trends are important elements of that blueprint. The formula is a product of old school, U.S.-style propaganda and has been used repeatedly in mass-marketed blockbusters. As others have shown, the use of a similar formula in the television coverage of the U.S. military's 2003 invasion of Iraq exploits time-tested wartime propaganda and high-tech multimedia spectacle (Giroux 2006a; Kellner 2005; Rampton and Stauber 2003; Rutherford 2004).

Part of the power of news programming lies in its ability to construct very specific ways of seeing and hearing the world in conflict. In that sense, war coverage on television operates no differently than everyday news. They both act as the "daily textbook for most Americans on what is happening in the world. . . . [The news] provides the public with an agenda of concerns . . . a vocabulary . . . and a sense of what dangers we face and from whom" (Dorman 2006, 13). Yet the magnitude of the events and the consequences of war commands special attention. Analyses of war coverage address its relation to historical fact, propaganda, and bias, but it is also important to position war coverage within the context of the industry that produces and distributes news content. To divorce televised war coverage from the entertainment industry is to decontextualize it in the most fundamental way.

This book explores how Cable News Network (CNN) and Fox News Channel positioned and packaged the U.S. military's invasion of Iraq in 2003 for a domestic audience. I place those two networks and the 2003 invasion of Iraq in the context of postclassical Hollywood filmmaking, one offshoot of which is high concept—a filmmaking practice inextricably linked to media conglomeration, new technologies, and an incessant self-preserving drive to market. When infused with critical theory, high concept is a valuable way to understand the politics of entertainment-driven war coverage. Although high concept typically describes a specific type of film, its value as a commercial tool has spread to television and even to a genre regarded as distinct from regular television fare: the news.

High concept provides an appropriate and useful framework for examining television news because contemporary television news shares industrial roots and ideological leanings with high-concept filmmaking. The corporate concentration of the U.S. film industry, which led to the development of high concept, eventually enveloped the television industry as well. Like high-concept films, television news is a commercial enterprise that is expected to turn a profit, and in the last thirty years, television news has absorbed the entertainment-driven commercial imperatives of its owners (Bagdikian 2000; McChesney 1999; Thussu 2003; Wittebols 2004). As a result, television news, like the film industry, conveys structural and ideological support for the system that enables it to prosper (Hallin 1986b; McNair 1998). Pro-business, pro-consumerism, and pro-military agendas therefore inform the construction of news, just as they affect the content of high-concept films (Wyatt 1994, 195–198).

The difference between the construction of news and the construction of films is that news ideally provides information that is essential to democracy. A healthy news system contributes to an informed citizenry. Yet in the commercial framework of high concept, news programming conflates citizens with consumers. In approaching their audience as a group to sell to and to be sold to advertisers, news producers adopt strategies that simplify and distort information about international and domestic events. If, as Herbert Gans argues, "journalism views itself as supporting and strengthening the roles of citizens in democracy" (2003, 21), then high-concept news exists to support and strengthen the role of news programming as a profitable commodity in a commercial television landscape. This book seeks to understand how television news got to this point and what the consequences of that development are.

I use high concept as an alternative to strictly quantitative analyses of news to challenge the notion that studies that focus solely on language are the most comprehensive approaches to unveiling the construction of meaning on television news. Additionally, I want to advance a more informed and systematic way of dealing with the idea that the representation of war on television is visually and aurally determined like a Hollywood film. To achieve this, I must strip television news of its stature as somehow divorced from and above the rest of television programming. I must also reinsert it into the entertainment industry and, using critical theory and aesthetic/formal analysis, examine the war coverage of CNN and Fox News Channel in relation to the history and contemporary state of that industry. In sum, I propose that studies that disregard or marginalize visuals, sound, and narrative and the industry that profits from the spectacular packaging of those three elements cannot fully capture the thrust of television news.

## Media, Spectacle, and Reality

The purpose of investigating the complexities of media and the politics of representation should be to problematize routine images and sounds. Images that reinforce each other without much opposition become routine and eventually become a natural part of the media landscape. In 24-hour war coverage, the slow pace of new information and the repetition of a small pool of shots naturalize the images and sounds of mayhem and military force. These types of shots are more prevalent in the beginning stages of war coverage. When commercial interruptions are absent and then limited, the concentrated coverage creates a visual and sonic diet of repetitive reports from journalists in the field. Reporters embedded with military units document events, but to say that such reporters "capture reality" betrays an ignorance of the process of mediation and of the labor-intensive and creative process of documentation.

The connection between documentary material such as news footage and realities such as war and death is unstable and subjective. Pointing the camera at one thing means that something else is not the object of our gaze; this is the most basic understanding of how certain needs, standards, or subjectivities select all taped or filmed events. However, more than simply pointing viewers in one direction or another, television news programming offers viewers *ways of perceiving the world* that are based on and mediated through images and sounds. By the third day of a war that is caught on video and broadcast via commercial U.S. media, images that were groundbreaking on Day 1 have become standard and the story of the war (complete with stars, characters, genres, and narrative trajectories) has emerged with clear narrators and the spectacle to support it.

State and corporate power occupy privileged positions on television and in news programming; we see this in the reliance of networks upon government and military sources as well as in their corporate parentage. By examining how the interests of the state and the media industries meet in the documentation of war, we can understand whose realities are represented and why some realities are excluded. Mainstream war coverage, as we presently know it, exists within an organized system of commercial media. In its capacity as television programming, war coverage in 2003 needed to accommodate commercial breaks and attract an audience to watch those advertisements. As a commercial product itself, war coverage—particularly coverage presented in concentrated form at the outbreak of war—presented a consumer-friendly vision of the war that did not deviate greatly from the way the state framed it.

The commercial- and state-friendly structure 24-hours news programming imposes on war coverage is spectacular in the Debordian sense. For Guy Debord (1967), "spectacle" is a catchall term for the mode of production as well as the product; it is capitalism, hegemony, and entertainment. The spectacle is "the existing order's uninterrupted dialogue about itself" (24). Debord characterizes the process of creating spectacle as the imposition of certain viewpoints and perceptions upon the populace through the process of mediation (18). Spectacle includes everything involved in the creation of the unreal in order to mask real, lived experiences and cloud perception. Debord's conceptualization of the spectacle was a radical one that dealt with absolutes. For instance, his conflation of entertainment with mindless consumption unequivocally implicates consumers in the hegemonic practice of the spectacle. He characterizes consumers of entertainment as passive, a position that ironically robs people of agency in a manner similar to that of the spectacle itself. Though not devoid of problems, Debord's delineation of three manifestations of the spectacle—information, propaganda, and advertisements—informs my understanding of the 2003 war coverage.

Henry Giroux recognizes the value of Debord's conceptualization of the spectacle, but he retools it in *Beyond the Spectacle of Terrorism* in order to emphasize the centrality of terrorism to the state and the media. Giroux lauds Debord for forcing us to recognize the links between power, acquiescence, technology, and capitalism (2006a, 38). Nevertheless, Debord's work addressed the escalating consumerism of the 1960s, not the confluence of war and media in the twenty-first century. Giroux remarks,

> Debord could not have imagined . . . how the second media revolution would play out, with its multiple producers, distributors, and consumers, or how the post-9/11 war on terrorism would transform the shift . . . from an emphasis on consumerism to an equally absorbing obsession with war and its politically regressive corollaries of fear, anxiety, and insecurity. (41)

Giroux uses the phrase "spectacle of terrorism" to mark the departure from a preoccupation with consumerism and signal the foremost concerns of the post–September 11 era. The political intentions that served Debord's work remain, but in applying the concept of spectacle to our present circumstances, Giroux highlights its historical flexibility.

"Different social formations produce spectacles unique to particular historical periods," Giroux writes, and the spectacle of the early twenty-first century situates us within "social relations constructed around fear and terror" (32, 30). Recognizing that state and corporate power do not restrict themselves to the promotion of consumerism, Giroux connects both entities through images: "The spectacle of terrorism conjures up its meaning largely through the power of images that grate against humane sensibilities. Rather than indulging a process of depoliticization by turning consuming into the only responsibility of citizenship, the spectacle of terrorism politicizes through a theatrics of fear and shock (30)." Giroux's use of the word "theatrics" recalls his claim that the spectacle of terrorism is about a "theatricality of power" and a "cinematic politics of the visceral" that have supplanted meaningful discourse (22, 24). Like Debord, Giroux places a premium on that which distracts us from and obscures the real. His allusions to "screen culture" and "visual politics" point to a hegemonic process of signification engendered by the tools of visual media and sustained by persuasive performance. Manipulation is key to the work of both Giroux and Debord; as the events on the screen are manipulated, they would say, so are we.

The ability to affect perception by manipulating events on screen was one of the first major advances in film form. Early Soviet filmmakers sought to manipulate reality and emotion through editing. Filmmakers like Lev Kuleshov and Sergei Eisenstein believed that the value of each shot was as important as the value of the juxtaposition of shots. Shots maintain those values in television news, but the context in which they circulate is much more complicated than in the era of silent film. Any given text of television war coverage stands alongside similar commercial and noncommercial texts. If all of the commercial news networks (and news programs) repeat the same stories and air similar shots, then the networks create their own context. They form what Edward Said (1997) calls a "communit[y] of interpretation" that defines itself according to its interpretation of a given topic and that has tremendous powers of influence (45–46). They define the parameters of the story, the mediated war, and "reality."

*The media that showcase the images and sounds of war without context or critique participate in and become the spectacle.* As a result, the connection between television news and reality is problematic and sometimes disastrous. Douglas Kellner (2005) notes the changes in war coverage since 1991: "Whereas in Gulf War I, CNN was the only network live in Baghdad and throughout the war framed the images, discourses and spectacle, there were more than 20 networks broadcasting in Baghdad for the 2003 Iraq war, including several Arab networks, and all of the television companies presented the war differently (64)." I would not go so far as to say that each company presented a wholly different version of the war, but each company did share one of the incubators of the news-reality connection: liveness. In *War and Cinema* (1984), Paul Virilio links manipulation to cinema and reality to television news, thus establishing a dichotomy that centers on immediacy and improvisation. Although he refutes this initial contention in *Desert Screen* (2002), his conflation of reality and liveness is common and ultimately harmful to a holistic analysis of television news. John Caldwell (1995) argues that the "theoretical obsession" with liveness overstates what is essentially television's false "claim to fame" (27, 31). What we should focus on, he writes, is what we see and hear or the style of programming. Caldwell does not consider how liveness can exist as an aesthetic of its own—as an element that operates in tandem with other stylistic attributes of a newscast.

Focusing on liveness as a marker of authenticity or reality, as Virilio does in his earlier work, may overstate its actual function, but it does not overstate the communicative power of the networks or how the news networks *perceive* the function of liveness. For Giroux, "Speed refigures our ability to access information, appropriate images from around the globe, and communicate with people on the other side of the planet," which underscores the "unprecedented power increasingly deployed by the news media" (2006a, 40). Immediacy is a key strategy employed by news networks to position their spectacle as hyper-real in a landscape of scripted and edited television programming.

## The Mediation of War

Producers of cultural artifacts attempt to confine meanings by establishing representational boundaries. The aesthetic of liveness reinforces these boundaries. If on the first day of the ground war in 2003 we see CNN's Walter Rodgers barreling through the desert in real time, we are transfixed by the image of sand and dust pouring out from the back of the vehicle in front of him as well as by his pseudo-poetic phrases that fetishized the U.S. military's "wave of steel" (CNN March 19, 2003). Immediately, CNN has framed the U.S. military's ground assault in terms of a none-too-subtle opposition between high technology and a barren sandy landscape. CNN's brand of representational control rests largely on the narrative it constructs, but we should also concern ourselves with how commercial news networks establish the parameters of reality and why.

What purpose does war coverage serve and how do strategies of representation—the mediation of war—figure into the larger aims? Looking back at organized efforts to capture war on film and video can help us begin to answer these questions. The "live" war coverage that we think of currently that features embedded reporters who insert themselves into the middle of pre-approved action is the descendant of war documentaries shot on film by another type of embedded observer. In World War II, the German army assigned one camera-man to every platoon; his job was to provide footage for and report to that unit's PK (Propaganda Company), which sorted and filtered information (Virilio 1989, 56). The Western Allies also had what Virilio calls "cine-commando units" at their service, and the Soviets had similar camera crews. Virilio argues that the purpose of these filmmaking ventures was not to transmit information to citizens. Rather, the reason for the "advanced mediation of military action" was strictly strategic and benefited the military (79). The most recent embedding system in the U.S.—and mainstream war coverage in general—did not depart from these earlier aims. Kellner does not mince words when he writes that the embedded reporters in 2003 "presented exultant and triumphant accounts that trumped any paid propagandist" (2005, 65).

Battles captured *on* film bear little resemblance to battles scripted *for* film, but the process of mediation influences both takes on the war. Film is probably the most compelling tool of persuasion and propaganda during wartime. The topic of propaganda and war may bring to mind the German actress and film-maker Leni Riefenstahl, whose state-supported film *Triumph of the Will* (1935) is the quintessential cinematic document of the Third Reich, yet the U.S. government and military engaged with the Hollywood film industry to achieve similar propagandistic aims. The oligopolistic organization of the filmmaking industry in Hollywood facilitated collusion with the government during World War II. The major studios were able to form a united front and oversee their collective interests while operating in the service of the war effort. Unlike radio's expropriation by the U.S. Navy during World War I, the motion picture industry was never under government or military control. However, because of the studios' financial ties to bankers and companies with interests in the military, the film industry at times became a mouthpiece for the U.S. war effort. The Hollywood studios were tied to the perpetrators, benefactors, and beneficiaries of war, and a similar—though not identical—set of circumstances enlisted the U.S. television industry in the service of the military during various conflicts.

As the television industry matured, its role in time of war differed from that of its predecessors; the state did not necessarily have to recruit it. If the Allies had cine-commando units during World War II, television broadcast networks boasted a volunteer force of telecommando units in subsequent wars. Many scholars have documented and analyzed the complicated relationship that television news outlets have maintained with the U.S. government during wartime (Andersen 1995; Andersen 2006; Carruthers 2000; Denton 1993; Der Derian 2001; Donovan and Scherer 1992; Englehardt 1994; Hallin 1986a, 1994; Herman and Chomsky 2002; Kellner 1992; Kellner 2005; Rutherford 2004; Shohat 1994; Taylor 1998). From Korea to Iraq and from outright censorship to the embedding system, that relationship has become more outwardly amicable but not any less problematic.

One hindrance to an agreeable relationship between the government and the media is the Vietnam syndrome—a term that describes the trauma the nation suffered as a result of the failure of the U.S. military to win the Vietnam War. The Pentagon's assertion that the news media somehow crippled the military's efforts in Vietnam plays a large part in this syndrome, but Daniel Hallin and Edward Herman and Noam Chomsky (2002) take issue with the Pentagon's argument. Hallin argues that the decreased support for the war in Vietnam was not the fault of the media but of "a political process of which the media were only one part." He reminds us that as casualties rose in the Korean War support for that war plummeted, even though "censorship was tight, and the World War II ethic of the journalist serving the war effort remained strong" (Hallin 1986a, 213). To blame the media for the U.S. military's loss in Vietnam not only oversimplifies the socio-political situation at the time, it also ignores the enormous role the television news media played in supporting the early stages of war, in acting as the mouthpieces of the Johnson and Nixon administrations, and in silencing the war's critics.

Both Kellner (1992) and Phillip M. Taylor blame the Vietnam syndrome for the Pentagon's adoption of the press pool system during the 1991 Persian Gulf War. Taylor writes that the Pentagon's fear of critical coverage ensured that journalists would never again have free access to battlefields (1998, xviii). He argues that in 1991 the military needed a quick and goal-oriented war, minimal attention paid to casualties, and the ability to exert total control over the media (3). To achieve these goals, the Pentagon resorted to a system that had been developing in theory ever since Vietnam but was not put into practice until the United States invaded Panama.[1]

When the U.S. military barred journalists from covering its invasion of Grenada in 1983, the result was a compromise between the military and the press known as the media pool system, which went into effect during the U.S. invasion of Panama in 1989 (ibid., 35). In the earliest conception of the pool system, journalists from different media would be on call in news pools, making it easier to evacuate them should war erupt. The military would filter the journalists' reports and the Pentagon would release them to the media. Journalists were not content with this method, and their uneasiness was understandable; in Panama, they had no access to the battlefield. They only had access to official news released by the Department of Defense. With war looming in 1990, the Pentagon released an initial plan for an updated pool system that contained a long list of restrictions and revealed a preference for television journalists. When the press (primarily print journalists) responded to the updated system with "a storm of protest," the Pentagon altered its initial plan by shortening the list of restrictions (37). Although a coalition of journalists filed a lawsuit against the pool system, for the most part the commercial mainstream media agreed to the revamped guidelines.

The pool system enforced in the 1991 Persian Gulf War was composed only of journalists from the United States, Britain, and France. These journalists,

---

1. A pool system was in effect during the Falklands War in 1982. Thirty British reporters accompanied the Royal Navy's Task Force, but Taylor writes that technical difficulties "severely hampered" the reporters' abilities "to report that conflict *while it was going on*" (1998, 36; italics in original).

who formed Media Reporting Teams, shared their reports with other journalists in Riyadh, but the military did not allow the press any unsupervised contact with the troops. All reports were subject to security review by the Public Affairs Office, and censorship was the norm. The Pentagon specifically prohibited "spontaneous interviews with troops in the field; the filming or photography of soldiers in 'agony or severe shock'; and the transmission of 'imagery of patients suffering from severe disfigurements'" (ibid., 35).

Sheldon Rampton and John Stauber argue that when the United States attacked Afghanistan after September 11, 2001, the Pentagon "realized that it had little to lose and everything to gain" by allowing journalists access to the troops (2003, 184). So began the embedding process, a restructured reporting system developed by Victoria Clarke, a former public relations agent who was the assistant secretary of defense for public affairs. In this new arrangement for war correspondents, the Pentagon embedded reporters with troops and allowed them to remain in that position if "they followed the rules," which prohibited independent travel and interviews off the record and mandated security reviews of reports (185). The mainstream media's compliance with the tenets of the pool system in 1991 led Taylor to argue that "military-media relations were in most respects more harmonious than they [had] been at any time since the Korean War" (1998, 9). Yet this arrangement seriously eroded the basic standards of journalistic independence; Melani McAlister writes that any "understanding of the anti war impact of television was fundamentally challenged by the relationship forged between the hyperrepresentation of Desert Storm and its joyful public embrace" (2005, 239). The 2003 system of embedded reporting, with its pretense of openness, was the pinnacle of the military-media relationship to date.[2] It succeeded for a while, but soon the relationship came under heavy critical fire. As William Dorman notes,

> Journalistic deference to Washington's official perspective is hardly a new phenomenon. In instances ranging from the Bay of Pigs to the Dominican Republic to Vietnam and the Iranian revolution, not to mention the invasions of Grenada and Panama, the first Gulf War, and the war against Yugoslavia, there was nothing unique about press behavior in 2003. What is new is that a

---

2. Reporter Lowell Bergman would disagree; he argues that "not since the Nixon administration has there been this level of hostility leveled at news organizations [by the government]" (quoted in Johnson 2007).

policy came to ruin so quickly that the president's assumptions and strategies as well as the record of how poorly the press had covered them simply could not be ignored. (Dorman 2006, 19)

Although the system did not hold, the initial preponderance of pro-military images and messages on corporate-owned television news indicates a confluence of military and industrial interests. Herbert Schiller argues that the economic prosperity and power of the United States has been largely a result of war. He writes, "A schizophrenic national condition has been created which finds mounting material well-being for large numbers of the population alongside of and largely dependent on omni-present and numbing terror" (1992, 75). Schiller notes that U.S.-led invasions that promote a certain brand of stability in a region do so in the service of commerce and maintains that the military and industrial sectors of the United States preserve an alliance that centers on information and communications technologies. "Special access to information," he writes, "justifiably has been recognized as a corridor to power" (77). Military-funded research and development produces technology that is available in the marketplace, and the benefit to both sectors of the military-industrial complex is clear. The military supports corporations through its contracts to develop technology for potential use in international conflicts, and corporations use their considerable weight to support defense spending and the goals of the military.

In between propaganda and converging interests lies consensus and nationalism. Edward Said writes that the complexity and diversity of the United States demands that the media construct "a more or less standardized common culture." The need to build consensus has long existed as "ideological rhetoric expressing a peculiarly American consciousness, identity, destiny, and role whose function has always been to incorporate as much of America's (and the world's) diversity as possible, and to reform it in a uniquely American way."[3] Said qualifies the result of this rhetoric as "the illusion, if not always the actuality, of consensus." According to Said, the media belong to "this essentially nationalist consensus" and operate in its service. Without manifesting as official mandates or outright censorship, consensus "sets limits and maintains pressures," thereby "drawing invisible lines beyond which a reporter or commentator does not feel

---

3. Said uses the term "America" to denote the United States of America and not the American continents.

it necessary to go" (Said 1997, 53). Consensus therefore can work to contain war coverage within invisible boundaries, which proceed to shape and contain perceptions of war.

While propaganda, converging interests, and consensus are potential answers to the question I posed at the beginning of this section, entertainment is another. Daya Kishan Thussu argues that as entertainment-oriented conglomerates have purchased news networks, television news has imported the formal characteristics of entertainment, resulting in a hybrid genre called infotainment (2003, 122). War coverage that falls under the heading of infotainment includes some of the visual aspects of high-concept news: high-tech reporting, complex graphics and satellite imagery, and a video-game aesthetic (117). This type of coverage results in a sanitized vision of war. The 1991 Persian Gulf War was a highly visible instance of this strategy.

Indeed, media critics and scholars have focused on the ways that the 1991 war coverage mimicked motion pictures. One *New York Times* writer referred to the 1991 coverage as "Iraq, the Movie" with "glamorous stars, non-stop virtual action and thus far not a single dead body on screen" (Taylor 1998, 46). Another commentator in 1991 likened the "Middle East theater of war" to "a European cinema of war" (ibid.). And writing about the air strikes during the 1991 Persian Gulf War, Baudrillard compares the prelude to a "Hollywood script," citing "the John Wayne language and bearing of the military spokesman" as a recognizable motif (1995, 2).

A wealth of on-screen evidence made a spectacle out of the war, and critics and radical scholars repeatedly compared the coverage of the war to filmmaking. The networks' use of stylized visuals in 1991 was one way that aesthetics decontextualized the war. Kellner argues that CNN's 1991 *Crisis in the Gulf* program is one example of how coverage "merged reports, military statistics, speculation, music and images of war into a nightly spectacle that normalized, and perhaps created a desire for, war" (1992, 88). Tom Engelhardt comments on the news networks' use of "extras," "vast sets," "preproduction," "pre-war teasers," "Star Wars-style graphics," "Disneyesque fireworks over Baghdad," and "post-1975 breakthroughs in Hollywood special effects" (1994, 84). Arthur E. Rowse notes the "glitz" of the news networks' logos and the push towards "newstainment" (2000, 29, 34), and Susan Moeller stresses the networks' need for "Hollywood caliber . . . snazzy production values" to keep people interested in war (1999, 19). Likewise, Rampton and Stauber emphasize the "engaging visuals" and use of night-vision shots in what they call an "Atari" war (2003, 179–180). McAlister writes that the 1991 war was a "staged media event" with a "new kind of media politics" at work (2005, 239), and Giroux describes how "air strikes . . . morphed into video games" (2006b, 3). Moeller argues that the networks needed style to make the biggest impact. She writes, "It is not the subject alone that makes the statement, it is the subject married to a technically proficient, stylishly appropriate packaging that reverberates in our memories" (1999, 42).

The 1991 coverage relied partly on eye-catching visuals, but excessive or overdetermined style without narrative cannot sustain a news cycle. Narrative is an essential tool for selling the motivations for war to the public and for selling war coverage to an audience, and a simplified narrative makes the sale that much easier. In 1991, the story reporters and politicians told the public was simple: the United States selflessly set out to liberate Kuwait from the invading forces of Iraq. Taylor writes that Saddam Hussein and the Iraqi army had to be constructed as a "serious military, economic, and ideological threat" so that ultimately the liberation of Kuwait would be "a war for the American way of life" (1998, 5). In order to make this case, the networks broadcast endless news stories about Iraq's brutality and potential nuclear weapons capabilities and about its capacity for launching terrorist attacks against the United States (Kellner 1992, 64). The U.S. government insisted that the war was against Hussein and "not the Iraqi people" (Taylor 1998, xix). The rescue narrative figured heavily in coverage of the 1991 war, in which the masculinized U.S. military rushed in to save the feminized Kuwait, which U.S. media characterized as a rape victim (Shohat 1994, 153).

Technologically and stylistically charged coverage exaggerated the operations of the U.S. military in 1991 and coexisted with a simple but compelling war narrative. The news networks used high-tech storytelling to sustain that drama, regardless of how detrimental that strategy was to critical inquiry. Kellner writes that before the war began, "the media helped the Bush administration by beating the war drums and producing an atmosphere where it was all too likely that military force would be used to resolve the crisis in the Gulf" (1992, 19). Lauren Rabinovitz and Susan Jeffords similarly claim that the promotion of the 1991 war in the media established a tone of acquiescence that discouraged any kind of critical interrogation about the war or its official justifications (1994, 12). Elements such as excessive style, simplistic narratives, and uncritical discourse operated in concert to deliver a new type of war coverage that achieved spectacular heights in 2003.

Propaganda, the convergence of corporate interests, and entertainment are accessible and valid answers to the question "What is the purpose of war coverage?" Each answer draws from a particular history and a unique set of circumstances, but if we take a step back and recontextualize the news, we can expand the scope of our answers. My analysis hinges not just on corporate ownership of media outlets or government pressure to promote a certain narrative but also on the network of relationships—political, economic, cultural, and aesthetic—within the contemporary media landscape. So while the infotainment label accurately characterizes a contemporary trend in television news, the term high concept properly contextualizes and broadens our understanding of the stylistic and commercial reach of television news.

## High-Concept Filmmaking

Theorists Siegfried Kracauer (1960) and André Bazin (1967) argued that photography and therefore motion pictures were capable of capturing and transmitting reality untainted by human involvement. For others such as Stanley Cavell, photographs are a "manufactured . . . image of the world" that "overcame subjectivity . . . by removing the human agent from the task of reproduction" (1979, 20, 23). Noël Carroll (1995) and Ella Shohat and Robert Stam (1994) argue that photos and films are representations, not reproductions of reality. The distinction is far from minor; the implications of representation resonate profoundly in any mediated text. The act of representing reality is an inherently compromised one. Shohat and Stam argue that "while on one level film is mimesis, representation, it is also utterance, an act of contextualized interlocution between socially situated producers and receivers" (1994, 180). Their point about film translates well to television news, a genre that subsists on its claims to authenticity, objectivity, and realism. Yet television news shapes perceptions of events through words, images, aesthetics, and its location within the media industries. We can hone our understanding of television news as representation and utterance—rather than as a reproduction of the real—by breaking through some artificial barriers that are specific to the medium and the industry. By removing news from its safe space as serious or educational television and locating it in the larger enterprise of entertainment media and by applying the film industry's notions and strategies of high concept to television news, we can see more clearly how images, stories, personalities, music, and marketing mediate international conflict.

The opening scene of Robert Altman's 1992 film *The Player* features a long take reminiscent of the opening of Orson Welles's 1958 film *Touch of Evil*. An intricate tracking shot offers the audience a view into the goings-on at a Hollywood movie studio. At one point, an older studio employee argues with a younger one about the merits of the long opening take from *Touch of Evil* in comparison to the opening shot of a more obscure British film. The self-conscious references snowball from there. Several times the camera allows viewers to spy into the office of one studio executive, Griffin Mill. Mill listens half-heartedly as screenwriters pitch him their story ideas. Buck Henry, playing himself, pitches a sequel to *The Graduate,* while two writers describe their film as *Out of Africa* meets *Pretty Woman* and another writer pitches a "psychic political thriller comedy ... with a heart." Though *The Player* does not mention the type of filmmaking the various pitches represent, all of the pitches in this scene from *The Player* were high concept. The entire concept for each film that was pitched was boiled down to one simple statement that referenced the equally concise and simplified concepts of other films.

The idea of high concept is easy to disparage because it represents the pinnacle of simplicity and commercialism. Contrary to the connotations of the word "high," the high-concept film is not designed to be cerebral in any respect. High-concept films are easy to pitch, easy to understand, easy to sell, and easy to consume. Although the practice of high-concept filmmaking epitomizes simplicity in moviemaking, when high-concept films are broken down into their component parts, they are as complicated as any other cultural product.

Justin Wyatt conceives of high concept as a spectrum based on "the look, the hook, and the book" (1994, 20). The look, or the style, attracts the eye. The hook, or the marketable concept and the marketing of that concept, lures the body into the theater seat. And the book, or the simple narrative, does not distract from the look. The look/image/style binds the film to the marketing, and synergy—made possible by corporate conglomeration—places that look in as many marketing venues as possible. The look, then, becomes interchangeable with the film, and the likelihood that the consumer will choose the film becomes that much more overdetermined.

Wyatt breaks down this summation of high concept into the specific attributes of these films that allow us to understand the films more completely. Style is central to the marketability of a high-concept film. For Wyatt, style is "the usage of techniques within the film that become characteristics of the film." Likewise, style is composed of "those elements within the film . . . which are central to the film's operation (and marketing)" (1994, 23). He lists six components of high-concept style: stars, character types, genres, simplified narratives, music, and the "look" of production design and cinematography. He is careful to note that not all components must be present in one film for it to be defined as a high-concept product.

Complex films are inefficient and high-concept films are easy. Ease of communication and understanding are necessary to ensure easy sales. All the elements of high-concept style are put in place to facilitate that ease. First, stars are economically efficient ways of communicating narrative information; they operate as shorthand. Second, because complicated characters and story elements are antithetical to the demands of high concept for efficiency, high-concept films use character types or personalities with only a few characteristics but with striking physical features. Third, high-concept films rely heavily on genre. Like stars, genres are recognizable and correspond to viewer expectations. Fourth, high-concept films emphasize strong relationships between visuals and music, allowing marketers to easily extract scenes or shots for use in music videos and other marketing endeavors. Equally significant for the purposes of marketing is the final element of the high-concept style: the look. Wyatt argues that the formal techniques of high-concept films stand out self-consciously. The more striking the visuals are, the easier it is for potential consumers to associate the marketing and merchandise with the film. Marketing and merchandising complete the high concept spectrum. High-concept films outpace all other Hollywood films in terms of their absolute connection to marketing and merchandising. This connection is key. These films cannot stand alone on their looks; the look must co-exist with an easily marketed and exploited concept (Wyatt 1994).

In high-concept films, the stars, characters, genres, narrative, music, and look are positioned diegetically and extradiegetically to foster enough consumer interest to warrant saturation releases. The mechanism is large and complex, but the magnitude of the high concept enterprise is fitting. There can be no small-scale high-concept films; those films would be simply blockbusters. The difference between the two lies in the degree to which the marketing and the film devote themselves to each other.

## High-Concept War Coverage

Although the promotion techniques media conglomerates use has fostered the existence of high-concept news, applying the model of high concept to television news and to war coverage specifically is not seamless. Not every high concept film incorporates every characteristic of high concept, and, in ways that I will address throughout the chapters, not every aspect of high concept is wholly translatable to television or television news. First, films and television programs are different commercial products. For example, their respective industries quantify their abilities to attract audiences differently. Furthermore, all of the commercial news networks cover the same stories at the same time, usually using similar footage. However, if anything, that circumstance actually escalates the importance of product differentiation—a vital facet of high concept. In fact, broadening the scope of high concept to include television encourages us to look at a genre like news as more than just informational programming or even infotainment. By analyzing how high concept works in television news, I offer an important paradigm for understanding how the news constructs the world around us.

My analysis problematizes the routine images, stories, and sounds on the war coverage of CNN and Fox News Channel during the first five days of the 2003 U.S. invasion of Iraq, from late in the day on March 19 to March 24. I have chosen CNN and Fox News Channel because they are the two highest-rated news networks on cable and because they project themselves as diametric opposites. I have chosen the first five days of the invasion as my sample for several reasons. First, the Bush administration initiated "decapitation" strikes on March 19, establishing that date as the official start of the invasion. Second, on March 19, CNN and Fox News Channel transitioned from typical news cycles and news programs to 24-hour war coverage. Third, that period of five days provided ample time for both networks to settle into their war coverage patterns. Fourth, CNN and Fox News Channel suspended commercial interruptions for most of the first two days of their war coverage—a significant decision that underscored the periodically uncomfortable relationship between advertisers and war. Finally, during this period of crisis, the viewing audience on both networks was substantially larger than usual, so the intensive viewing of repetitive images and stories and speculative commentary was a common experience.[4]

After viewing and transcribing approximately 200 hours of CNN and Fox News Channel war coverage, I divided the coverage into five main aspects: graphics/animation, titles and text crawls, commercial breaks, coverage by studio journalists, and commentary by military and other experts. I then broke down the coverage by studio personnel into ninety-eight themes and topics that appeared throughout the coverage. That enabled me to count the occurrences of specific topics and, more important, to deconstruct and reconstruct the war

---

4. Recent historical precedent shows that the first days of a crisis provoke the highest news viewership. For instance, in the first four days after the September 11, 2001, terrorist attacks, CNN attracted nine times its average viewership, while Fox News Channel attracted five times its average viewership (Project for Excellence in Journalism 2004, "Cable Audience" section). Likewise, in March 2003, Fox News Channel's ratings increased by 50 percent, and CNN's ratings increased by 44 percent (Romano 2003).

narrative of the networks. This process enabled me to determine the extent to which the war narrative drew from the government's narrative and the extent to which it was constructed by the networks.

In addition to performing textual analysis of coverage of the war, I use theories of film, television, news, stars, characters, genre, and narrative to apply Wyatt's theory of high concept to television news. I also apply critical theory and scholarship focusing on power, hegemony, and the production of media under capitalism by theorists and scholars such as Guy Debord, Henry A. Giroux, Edward Said, Paul Virilio, Daniel Hallin, Robert McChesney, and John Fiske.

The organization of this book corresponds to each branch of high concept: narrative, stars and genres, character types, visuals and sounds, and marketing and merchandising. However, before I can approach those specific areas, it is necessary to connect the development of television news to the development of high concept.

In chapter 1, I explain how high concept production techniques have expanded beyond the filmmaking of the 1980s and 1990s into the television industry and news programming. The development of high concept depended as much on the film industry's shifting organization as it did on changing distribution patterns and innovative marketing techniques. The patterns of conglomeration that fostered a synergistic marketing environment for high concept have brought television and news programming into the same corporate families as film production companies. That shared parentage is one reason for the link between television programming and high-concept filmmaking. High-concept productions and corporate concentration both rely on highly commercialized products and heavy marketing to increase consumption. After tracing the development of news programming and providing background on CNN and Fox News Channel, I move forward to the specifics of high-concept war coverage in 2003.

In chapter 2, I assess the construction of a particular war narrative on CNN and Fox News Channel. I achieve this by examining the original narrative—disseminated through Department of Defense briefings—and by tracing how journalists at CNN and Fox adopted, modified, and transmitted this message. The original story was one of revenge that established the September 11, 2001, terrorist attacks as the cause and the U.S. invasion of Iraq as the effect. The Bush administration claimed that the old nemesis of the United States, Saddam Hussein, was associated with the terrorist organization responsible for the September 11 attacks. In this narrative, Hussein's alleged stockpile of weapons of mass destruction (WMDs) and ties to Al-Qaeda made another attack on the

United States inevitable. In order to protect the citizens of the United States and innocent civilians in Iraq, the U.S. military needed to intervene. In other words, a simple series of cause-and-effect events let to war.

Simplicity characterizes the war narrative I analyze in this chapter. Simple narratives are integral components of high concept, and they are vital to the type of war coverage promoted and presented by CNN and Fox News Channel. Equally important were the ways that CNN and Fox News Channel personnel initiated and set the parameters of the war narrative. These were significant both in terms of the narrative and in political terms, particularly because the events of the war consistently proved difficult to contain. That difficulty is apparent in the six events I analyze in chapter 2: the "decapitation" strikes that opened the war on March 19; the "shock and awe" bombing campaign that followed; the videotape of Iraqi celebrations at Safwan; the raising of the U.S. flag at Umm Qasr; the speculation surrounding the discovery of a suspected chemical weapons plant at Al-Najaf; and the capture and videotaping of U.S. prisoners of war (POWs). The different ways that CNN and Fox News Channel represented the war's events indicated different levels of commitment (but commitment nonetheless) to the Bush administration's vision of the war and to the commercial viability of the coverage.

In chapters 3 and 4, I examine how formula and intertextuality were important in efforts to maintain the reductive war narrative. Formula and intertextuality were most evident in the two networks' use of and/or references to past wars, film genres, movie stars, and established character types. Here I discuss not only the genres implied by the language of studio journalists and commentators but also their specific references to Hollywood formulas. The stars of the war were not limited to President Bush and Saddam Hussein. The war narrative the networks followed included a number of other key faces, including Iraqi information minister Mohammed Said al-Sahhaf and the networks' own embedded reporters. It also included standard character types—villains, heroes, and false heroes—that, for the most part, originated in the two Bush administrations. These character types supplemented the war narrative, supported the philosophy of the war perpetuated by the Bush administration, and were eminently marketable by virtue of their simplicity.

Chapter 5 focuses on the high-concept look and sound of war coverage at CNN and Fox News Channel. The coexistence of certain kinds of images and sounds exemplify the excessive style of war coverage at these networks and the predominance of the marketable concept of the war, which I characterize as altruistic revenge the United States took on behalf of Iraqis and September 11 victims that was achieved through technological superiority. This marketable concept is communicated clearly in the production-related elements of the coverage.

Marketing and merchandising are two fundamental aspects of high concept. In chapter 6, I argue that CNN and Fox News Channel plainly used coverage of the 2003 war to advance their commercial aims by adhering closely to the war's marketable concept. Armed with this concept and a simple story spread by the Bush administration and the Department of Defense, the two networks marketed the war narrative for commercial ends. One way they achieved this was by incorporating the style of advertisements to promote their war coverage. In these promotional spots, visuals and iconography represented and endorsed the marketable concept. They also advertised network-produced merchandise, merchandise produced by other members of the networks' parent megacorporations, and merchandise produced by companies outside the parent corporations that tied into and helped promote the marketable concept of the war. Advertisements for all three types of merchandise played a large role in promoting the war narrative in the commercial sphere. The result was the full-scale commodification of a simplified version of events.

The conclusion examines several questions that have arisen from my work. I consider the value of high concept to the field of media studies and I pick apart common misperceptions about news as a genre. I also address questions relating to the international and domestic scope of high-concept news as well as its effects on news viewers. I conclude with a discussion of the noticeable absence of war coverage from news programming and some of the other genres and media that have represented the war.

What follows is a meticulous examination of the first five days of the 2003 war coverage. The analysis is not exhaustive. However, by focusing on the details of that first week of the invasion, I encountered a number of large issues—some expected and some completely unexpected—that support my argument that the practical and political aspects of high concept are thriving in commercial television news.

# 1

## *High Concept, Media Conglomeration, and Commercial News*

Privately owned journalistic enterprises, like commercial newspapers and television news outlets, are participants in a "crowded information marketplace" (McNair 1998, 101). Consumers who purchase a newspaper or choose to watch a certain news program (paying for it by watching the accompanying advertisements and subscribing to a cable service) have passed over other options. Commercial journalism must have use value and exchange value because it vies for audiences and depends on them for revenue. In Brian McNair's words, news "must be both functional and desirable" (ibid.). News whose content and format appeals to audiences sells, and if the result is inexpensive for the producer, all the better.

The circumstances surrounding commercial journalism are more intricate than the simple quest for profit, however; it must also perpetuate the system that sustains it. William Dorman writes that theoretically "a privately owned press unrestrained by government provides for a free marketplace of ideas that makes it possible for citizens . . . to debate alternatives, become aware of abuses of state power, and ultimately, hold governments accountable" (2006, 13). However, according to Brian McNair, the owners of "lucrative capitalist enterprises" like media conglomerates "understandably . . . use their media to support [capitalist] economies and to preserve socio-political systems which allow them to go on generating profits" (1998, 103). The point of the commercialization of journalism, then, is to create an environment in which it is profitable to continue commercializing journalism.

Consequently, the function of news produced by media conglomerates is intimately aligned with the worlds of commerce and entertainment. James Wittebols (2004) argues that the result of increased media conglomeration is that most television genres, including news, have incorporated the narrative form of soap operas in order to make those particular genres profitable. Daya Kishan Thussu also asserts that entertainment values are so pervasive within media conglomerates that television news has become infotainment (2003, 122). The interest of giant media corporations in maintaining the news as a profit center, asserts Robert McChesney, has led to the monopolization of news time and space by "inexpensive and easy" stories (1999, 54). Richard Cohen, a former *CBS Evening News* producer, likewise argues that what is important in news has given way to what sells (1997, 39). In short, the need to make money from television news has diluted its substance.

These critics argue that conglomeration of media corporations has removed something from television news. I argue that the gap has been filled by a multi-faceted mode of newsmaking that borrows its marketing strategies from high-concept Hollywood films. High-concept marketing and merchandising owe their expansion to the rise of media conglomerates, and television news is part of that same industrial development. To believe otherwise would be to separate news programming from its context in the media industry and its home in television.

This chapter traces the history of the film and television industries in order to draw out the connections between high concept, the contemporary state of media conglomeration, and the place of news programming within media conglomerates. My aim is to examine how news programming came to assume the role of infotainment and adopt the strategies of high concept. I argue that the uniqueness of contemporary commercial television news is bound to the relatively recent large-scale convergence of the media industries, which has facilitated the union of high-concept values and journalistic values.

## High Concept: Structural Roots

In *High Concept,* Justin Wyatt debunks the popular definition of high concept in favor of a more thorough understanding of the term. He begins with its definitions within the television and film industries, all of which emphasize the concept that simple ideas in films lend themselves to successful marketing. He then connects such definitions to a larger context in which high concept represents just one route that mainstream U.S. film has taken since the disintegration of the classical Hollywood studio system, the route of product differentiation. In a market flooded with sameness, product differentiation is achieved through the style of the film and the way in which the film and its marketing coalesce. Wyatt insists that the combination of aggressive marketing, merchandising, and style makes high concept a "more exact category of differentiation" from typical Hollywood films (1994, 105). Wyatt's definition of high concept emphasizes the connection between the market and the product:

> High concept can be described . . . as one strain of contemporary American cinema whose style has a direct economic motive. This economic strategy depends on conceiving the market for film as splintered through product differentiation and market segmentation. (108)

In other words, high-concept films are successful because the film industry is able to bring unique products to specific target audiences. It is necessary to look back to the founding of the film industry to understand how U.S. film got to this point—and how U.S. television news has followed.

High concept is indebted to the material development of the U.S. motion picture industry (even though Wyatt officially attaches high concept to post-classical Hollywood cinema). This development takes the form of equipment, funding, and the structures put in place to expedite filmmaking for commercial purposes. Hollywood began in an organized fashion with the Motion Picture Patents Company (MPPC), a patent pool orchestrated by the Edison Film Company and the American Mutoscope and Biograph Company in 1908 as an attempt to halt ongoing litigation over patents (Anderson 1985, 138). Following the anti-competitive example of companies such as Standard Oil and General Electric, eleven filmmaking companies cooperated to monopolize ownership of film stock and cameras and the production, distribution, and exhibition of films. This action organized and standardized the fledgling U.S. film industry, but the MPPC's methods resulted in an antitrust suit brought by William Fox. In 1915, a U.S. District Court determined that the MPPC was operating illegally, and in 1918 the MPPC's cause was "officially terminated" when its appeal was denied (147).

As one monopoly faded away, ambitious men forged an oligopoly to take its place. In the late 1910s and into the 1920s, the "Big Three"—First National, Paramount-Publix, and Loew's/MGM—forged a successful vertical integration in Hollywood when they each acquired production, distribution, and exhibition facilities. However, the development of sound technology and the consequences of the Great Depression disrupted the organizational structure of the 1920s. RKO emerged from RCA's involvement in sound technology, and Warner Bros. grew when it acquired First National from Fox (Thompson and Bordwell 1994,

234). From 1930 to 1945, Hollywood boasted a collection of five vertically integrated studios, otherwise known as the "Big Five," which included Paramount, Loew's/MGM, Fox (renamed 20th Century-Fox in 1935), Warner Brothers, and RKO. The studio system survived the Great Depression, religious censorship, celebrity scandals, and the House Un-American Activities Committee, but ultimately, the U.S. Justice Department proved to be its undoing. In 1948, as a result of the ruling in the second antitrust suit against the motion picture industry, the five vertically integrated studios were required to sell either their production facilities, their distribution facilities, or their exhibition facilities in order to comply with antitrust law. The Big Five elected to part with their exhibition venues. After that, the studio system's decline was a slow one; the breakup of the oligopoly took ten years to complete.

The new structure that developed after the breakup led directly to the beginnings of high concept. Though the studios leased their spaces to independent producers who assembled productions for distribution, income from leasing did little to counterbalance the loss of income as the result of decreasing theater attendance. In 1955, film attendance reached a low of approximately 2.3 billion patrons per year; it dropped even lower in the 1960s to approximately 1 billion (Cook 1996, 512; Thompson and Bordwell 1994, 697). Some in Hollywood held television (the "convenient stock villain") responsible for the lethargic attendance (Anderson 1994, 2). With annual receipts totaling an all-time low of $900 million in 1962, Hollywood looked to spectacles and musicals, hoping these films would reverse their financial decline (Cook 919). When such films as *Doctor Doolittle* (1967) and *Star!* (1968) flopped, the trend halted. In addition, television networks began producing their own films, increasing the supply of films released and bidding very high for talent, all of which resulted in high production budgets. In 1968, the networks also stopped buying films to air, and the market was flooded with the films of more and more producer-distributors (Londoner 1985, 605–607). The studios felt imperiled, and they believed that their ability to offset losses from flops with revenue from hits was threatened. Wyatt traces the origins of high concept to this moment. The studios saw no room in their businesses for films that did not guarantee a return on investment (Wyatt 1994). The less risk involved in producing and marketing films, the better.

The first wave of mergers in the 1960s was one of the primary components of the producer-distributors' new focus upon bankable films. The four major film distributors—Universal, Paramount, Warner Brothers, and UA—either merged with conglomerates or felt the conglomerates' aggressive advances. Wyatt notes two significant consequences of conglomeration that fostered high concept. First, conglomerates emphasized the long-term stability of the industry (Wyatt 1994). Second, conglomeration made it possible for each film to be exploited in any of the company's sectors; this synergy proved to be more valuable than the exhibition outlets the studios had relinquished. Conglomeration transformed the way companies profited from films and focused attention on the most profitable aspects of Hollywood films.

In its quest for higher profits, Hollywood began overproducing and lost $500 million between 1969 and 1972. Banks began intervening to control costs, and producers scrambled for something new to attract patrons (Thompson and Bordwell 1994). Ironically, the novelty they were searching for arose from their own organizational misfortunes. Beginning in the 1930s, the MPPDA voluntarily applied its own censorship system in an effort to avoid government regulation. The Production Code Administration rigorously policed films to make sure that the content would not offend audiences. Films not passed by the PCA could not be exhibited in studio-owned theaters; these films were therefore denied the benefits of major theaters in major markets. This institutional censorship collapsed when the major studios had to sell off their theaters. There was simply no mechanism for punishment. And in 1968 the Motion Picture Association of America instituted its ratings system. Relaxed standards permitted liberal content, and filmmakers from 1969 to 1975 had some room to push against the boundaries that had defined (and confined) classical Hollywood cinema.

This strategy was successful only to the extent that targeting a niche can succeed, and media conglomerates began seeking the big money of blockbusters to minimize their risk. Blockbusters were Hollywood's "return to large-scale, grand filmmaking" on the scale of the epics and musicals of the 1960s (Wyatt 1994, 77). For Wyatt, the typical blockbuster is "a pre-sold property (such as a best-selling novel or play), within a traditional film genre, usually supported by bankable stars (operating within their particular genre) and director" (78). Because of these attributes, blockbusters had a "built-in audience appeal" that helped film distributors court vital foreign markets (80). Consumers in these markets were not interested in personal stories; they wanted action-adventure films. Studios reduced the number of films they released in order to transfer funds and energy toward the new blockbusters.

High concept stems directly from the revival of the blockbuster and other industrial circumstances, including conglomeration, the rise of television, new technologies, and improved film distribution strategies. The profitability of new television technologies such as cable, pay television, and home video enticed the film industry to establish itself in these markets. In fact, Wyatt argues that the strongest conglomerates of the 1980s owed much of their success to "diversification and innovation into other delivery systems" (1994, 84). New television technologies assured new exhibition windows as well as tremendous marketing possibilities. Another new strategy associated with post-classical Hollywood was the saturation release: following the lead of low-budget film companies such as American International Pictures, major studios began to release films in hundreds of theaters on the same day (Thompson and Bordwell 1994). Wyatt notes how well suited this strategy was to high-concept films. If a saturation release is to be successful, the studio needs to generate a high level of positive awareness of the film prior to its premiere with aggressive marketing across a variety of advertising platforms (Wyatt 1994). Television was a prime location for the ad campaigns that preceded saturation releases. Everything was in place for high concept. The ways conglomerates used their resources to market high-concept films were sophisticated, even if the films themselves were the antithesis of sophistication.

## Conglomeration: The Economic, Regulatory, and Industrial Roots of High-Concept News

When *High Concept* was published, Wyatt predicted that although high-concept filmmaking might gradually slow down because of demographic changes and the escalating price of filmmaking, it would not collapse. Rather, he believed that high concept would "continue, to some extent, into the future" (Wyatt 1994, 190). Indeed, the economic and industry-specific circumstances that helped generate high-concept films in Hollywood in the late 1970s and 1980s remain intact, although they are now more centralized. The possibilities for marketing and merchandising have multiplied with the exponential growth in the size of conglomerates, not to mention the decrease in their actual number. The history of conglomeration and the effect of corporate ownership on journalism provide a foundation for understanding the place of CNN and Fox News Channel in the commercial television news landscape.

### *Media Organization and the State*

Media conglomeration in the United States developed in the context of a legislative atmosphere that favored corporate control of communications technology, in spite of the fact that some elements of that technology (e.g., the airwaves) are public property. For instance, the Communications Act of 1934 left the structure of the communications industry intact, even though it was already dominated by patent pools and corporate agreements. Robert Horwitz notes that the so-called natural monopolies (AT&T and Western Union) continued unabated, aided by regulation that protected them from potential competitors (Horwitz 1989, 124). The 1934 act reinforced the three basic technological distinctions that had been constructed in the 1920s: "print, wired electrical communications, and broadcast communications" (ibid.). Horwitz argues that these distinctions served democratic ends because they separated content producers from distributors. In this tripartite system, each communications platform was handled by a separate corporation and there was no vertical integration in the communications industry.

The film industry remained outside of the tripartite structure of the U.S. media industry, but film's relationship with broadcasting persisted. Christopher Anderson notes that as the communications industry expanded in the 1920s, the film industry attempted to diversify into broadcasting in order to promote its products and determine the future of television. Warner Bros. acquired two radio stations in the mid-1920s and RCA created the movie studio RKO in 1928. Paramount bought 49 percent of the CBS radio network in 1929. However, financial upheaval during the Great Depression forced CBS and Paramount to separate and RCA to stop financing RKO. In 1936, Warner Bros. tried to buy into a radio network but failed. The studio recognized that distribution was the key to power in the radio industry, so without a network it would be pointless to linger in radio. Nevertheless, in the 1930s, the film industry and the radio industry constantly exchanged talent and content (Anderson 1994).

Radio did not prove to be the promotional powerhouse the studios were hoping for, and Hollywood virtually abandoned that medium as the 1930s drew to a close and began to turn its attention to the new medium of television. In that decade, studio contracts with independent producers included the rights to future television broadcasts. Anderson writes that in the wake of the 1948 U.S. Supreme Court ruling that forced studios to sell off one arm of their holdings in order to comply with antitrust laws, Warner Bros. tried to "salvage aspects of the studio system by integrating film and television production." However, after failing to incorporate radio into their holdings and then being forced by the Court to divest themselves of their exhibition holdings (the theaters they owned), the studios discovered that the 1948 antitrust ruling practically ensured that they would not be diversifying their holdings with television companies either. The 1934 Communications Act gave the Federal Communications Commission (FCC) the power to deny a broadcasting license to those who violated antitrust laws. When the FCC investigated the studios' eligibility for broadcast licenses in 1948, it initiated a freeze on new licenses because the existing channels had been placed too close to each other, resulting in interference. The license freeze, which lasted from 1948 to 1952, barred the film industry from expanding into television, during a period when radio networks, which had not violated antitrust laws, were able to buy up an increasing number of television stations (Anderson 1994, 32–33, 39–40).

*Media Mergers and Acquisitions*

In spite of the rules put in place by Congress and the U.S. Supreme Court to prevent motion picture studios from becoming monopolies, the first wave of mergers in Hollywood took place in the 1960s, just over ten years after the studios were forced to break apart. This wave of mergers took place primarily to stabilize the industry and spread risk through the creation of various "profit centers" (Balio 1998, 61). The wave consisted of three different organizational maneuvers: non-entertainment corporations acquired movie companies, entertainment conglomerates acquired movie companies, and movie companies diversified on their own to become conglomerates. In 1962, Music Corporation of America (MCA) bought Universal Pictures; in 1967, Seven Arts bought Warner Bros. and Transamerica Corporation bought United Artists; in 1968, Kirk Kerkorian purchased and almost entirely dismantled MGM; and in 1969, Warner Communications formed as a result of the merger between Warner Bros.-Seven Arts and Kinney National Services (Thompson and Bordwell 1994). Tino Balio points to Gulf & Western, which bought Paramount Pictures in 1966, as the "quintessential sixties conglomerate" because of its transindustrial holdings, including fertilizer and real estate holdings (1998, 61).

A second wave of media mergers in the 1980s signaled another attempt by the studios to vertically integrate so they could maximize their ability to advertise across platforms and strengthen distribution (ibid.). However, in order to increase the number of holdings under one parent company, the major studios pushed the boundaries of antitrust legislation once more. Under the Reagan administration, the Justice Department did little to prosecute antitrust violations (McChesney 1999, 311). The Reagan administration's lax attitude created an environment in which the major studios could begin buying up theaters again without fear of legal consequences. Thus the major studios reacquired the exhibition arm of their holdings that the Supreme Court had specifically dismantled over three decades earlier. Columbia Pictures commenced reintegration by buying some New York City theaters in 1986, and MCA, Paramount, and Warner Bros. followed suit shortly thereafter (Balio 1998). The merger wave continued throughout the 1980s. In 1981, Transamerica Corporation sold United Artists to MGM and Marvin Davis bought 20th Century-Fox; in 1982, Coca-Cola bought Columbia; in 1985, Ted Turner bought MGM (only to resell some of its assets later) and Rupert Murdoch bought 20th Century-Fox from Davis; and in 1989,

Coca-Cola sold Columbia to Sony and Time, Inc. bought Warner Communications (Thompson and Bordwell 1994). By 1989, the number of media-related mergers reached 414 (Schatz 1997).

The early 1990s saw more major mergers: Matsushita Electric Industrial Company purchased MCA, and Pathé Communications bought MGM in 1990 (only to lose it to Crédit Lyonnais in 1992) (Thompson and Bordwell 1994). After a lull, the media industry mergers continued in the mid-1990s, when conglomerates such as Disney and Westinghouse sought to acquire over-the-air television networks. In 1995, the FCC repealed the finance and syndication rules that had prevented the networks from having a financial stake in their programs beyond their initial airings and from creating syndication companies. Liberated from these restrictions and spurred on by the synergistic potential of wholly owned television programs, the conglomerates eagerly purchased the networks. General Electric purchased NBC in 1986 when it acquired RCA, and in 1995 (and within days of each other), Disney bought ABC/Capital Cities and Westinghouse acquired CBS (Balio 1998). That year, the total number of mergers in the media industries came to 644 (Schatz 1997).

The following year, Congress passed legislation that revised the 1934 Communications Act in a way that made greater concentration of ownership possible. Although it is known mostly for imposing a voluntary television ratings system and mandating the placement of V-chips in new television sets, the 1996 act significantly relaxed ownership regulations. Those who promoted the Telecommunications Act of 1996 believed that "the public interest is best served by a reduction of regulatory control" and that the free market would efficiently regulate communications technologies and services on its own (Parsons and Frieden 1998, 13). The antitrust laws that seemed to stand between fair and unfair competition increasingly lost traction in the corporate-friendly environment that led to the 1996 act.

Less than forty years after the dissolution of the studio system, the major studios were able to once again acquire the movie theaters that distributed their products. In addition, television networks sat alongside movie companies as holdings in conglomerates.

### Television Journalism and Media Conglomeration

As journalism became a profession, its practitioners moved toward objectivity and away from the type of bias that had characterized newspaper reporting in the early 1900s. Although this move was initiated for pragmatic reasons—that is, to keep up an appearance of legitimacy and avoid alienating readers and advertisers—journalists represented themselves as working for democracy and the public (McChesney 1999). Journalism became the watchdog profession that kept the powerful in line and was not controlled by state or private interests (McNair 1998, 83). McChesney writes that the role of journalism in commercial media has been one of public service, at least superficially. The commitment of media corporations to the public interest through journalism is one of the primary justifications they have used to gain protection under the First Amendment (McChesney 1999, 49). As an objective democratic institution, news programming became a way for broadcast television networks to operate in the public interest and fulfill the public service mandate that station licenses carry (Wittebols 2004, 67).

Over the course of thirty years, that mandate was subverted as broadcast network news underwent fundamental format changes due to cost cutting and ownership shifts. In 1948 and 1949, CBS and NBC, respectively, aired newscasts that lasted a total of fifteen minutes; by 1963, both networks were using a half-hour newscast format, even though network news operated at a deficit until the late 1970s and early 1980s. Networks offset these losses with more profitable programming. Network news also benefited from the financial success of local news operations, which were the antithesis of network-style reporting. Local news broadcasts, spurred on by the counsel of marketing consultants, pursued emotional, violent, and sensational stories that sharply contrasted with the staid and measured professionalism of news anchor Walter Cronkite at *CBS Evening News.* When Cronkite left CBS in 1981, ratings for *CBS Evening News* declined. Because ratings determine the price that networks can charge advertisers for commercial time, the decline was the impetus for a number of changes (Wittebols 2004).

Van Gordon Sauter, the man brought in to restore the CBS newscast to its former ratings dominance, came from a local news background and implemented the style that had been so profitable for him in that venue (Wittebols 2004). Sauter, whose tenure lasted from 1982 to 1986, was not a proponent of the split between news and entertainment advocated by CBS in the 1960s and 1970s (Glynn 2000). The philosophy at CBS News in the Walter Cronkite era was one of "professionalism and news sense," not entertainment (Wittebols 2004, 68). For example, the 1976 CBS Standards Handbook prohibited the use of music, reenactments, or any tactics of a "show biz" nature in the news (ibid.). In contrast, under the tutelage of marketing experts, local news producers privileged style over substance and began the practice of connecting news stories to the topics of prime-time dramatic series. Sauter brought that style of news production to the network.

ABC had already implemented a narrative-driven news format when Sauter was hired at CBS. ABC brought Roone Arledge from *Wide World of Sports* and *Monday Night Football* to ABC News in 1977. As president of ABC News, Arledge was there to "jazz up" the news, and he did so by importing the idea that had worked so well for him in sports programming: the story line. Arledge's appreciation of narrative and drama resulted in a "fusion of entertainment values with the news." His primary innovation was the implementation of an entertaining narrative format in newscasts (Wittebols 2004, 75–76). By the mid-1980s, both CBS and ABC had largely abandoned their obligation to provide a public service and had embraced the values of the entertainment industry in an effort to push up ratings.

Conglomerates also changed the focus of news. In 1995, fearing that a *60 Minutes* story about the links between tobacco and cancer would generate a multibillion-dollar lawsuit by tobacco producer Philip Morris Corporation that might jeopardize Westinghouse Electric's proposed merger with CBS, the network postponed the story (73). The "news" became less about fulfilling a public service and more about protecting the assets of the corporation. And when Viacom bought CBS in 1999, fears arose that the conglomerate would use CBS News as a synergistic asset. Indeed, one instance of cross-fertilization involved a drug awareness news program that aired on CBS and MTV, another of Viacom's holdings (Wittebols 2004, 73). When Disney purchased ABC, news programming regularly promoted Disney films, citing "justified journalistic interest in the films" (Sterngold 1998). These are just a few examples of the influence that conglomeration has had on television news, but they illustrate the many ways in which news is not so different from other television programs. The news is vulnerable to the demands of corporate parents and its content can be shaped to meet these demands.

The transformation of network news from loss-incurring reporting to profit-generating programming produced Ben Bagdikian's assessment that the news is now "a handmaiden of its owners' corporate ambitions"—an indictment that acknowledges that in the United States, there is little room in today's newsrooms for public service or the promotion of democracy (2000, xi–xii). Many developments have eroded the networks' commitment to public service, including the FCC's "deregulation frenzy" in the 1980s (Wittebols 2004, 83). In his pursuit of Reagan-era deregulation, FCC chair Mark Fowler suspended the Fairness Doctrine, which ensured that broadcast stations gave air time to both sides of controversial issues. He also suspended limits on advertising time and quotas for public-interest programming. Thus, the networks could do what they wished with unprofitable programming. In Wittebols's estimation, deregulation, cutbacks, marketing consultants, and corporate control have moved the news away from public service and toward profit- and entertainment-oriented programming.

The belief that corporate ownership of news outlets has interfered with the integrity of news content is widespread for several reasons. In addition to a near-absence of news reports that are critical of corporations and the "national security state" and a disproportionate reliance on official sources, conglomerate-owned news outlets also perform public relations work for their corporate siblings (McChesney 60, 53). Richard Cohen views this high degree of corporate control as a crisis and argues, "The more market driven news becomes, the greater its determination not to rock the boat." He particularly bemoans the state of television news, in which "entertainment values and the irrelevant" sell a lifestyle that advocates a corporate worldview and resists critical interrogation of existing power structures (1997, 33–34). The insertion of corporate agendas in news content is an important element in the efforts of corporations to sustain the system that sustains them.

McNair argues that owners of media companies wield cultural power through their products (McNair 1998, 102–103). McChesney similarly holds corporate ownership accountable for the contemporary state of journalism in the United States, but he also blames the "rightward movement" of U.S. culture and points to the proliferation of conservative guests on news programs and the rise of conservative media mogul Rupert Murdoch to illustrate his point (1999, 62). Murdoch's newspapers have a reputation for mirroring his personal brand of right-wing conservatism because his staffs are either like-minded or they learn to be (McNair 1998, 107).

Daniel Hallin posits that the dominance of Reaganism greatly impacted the political leanings of the news. He writes that the patriotism typical of Reaganism makes sense on television because news networks win more viewers with pro-U.S. sentiment than with critical reporting. Television news anchors "appeal for 'votes'" daily, and they achieve victory more easily if the anchors "wrap [themselves] in the flag and praise the wisdom of the People" instead of dealing with controversy (1986b, 38–39). However, ideology is not the only influence on journalistic output. The market is also extremely important. Commercial media corporations need audiences and preferably audiences of "quantity and quality." If we look again to Rupert Murdoch, whose control over the content of his media outlets is well documented, one example of the market's dominance over ideology stands out. The British satellite network Sky News is Murdoch's prestige news outlet; it represents his attempt to "gain some respect as a broadcaster" (McNair 1998, 110). He gave the network's journalists relative autonomy and succeeded in attracting a "quality audience," or an audience of educated people with disposable income. As a result, Sky News was lauded for its content and Murdoch was able to establish a reputation as a serious newsmaker and pursue his goal of creating a media empire. When pursuing a certain audience, the owner's economic interests are most important. A *New York Times* op-ed piece argued that if Murdoch was a "right-wing ideologue . . . he'd be much less dangerous." His inherent danger lies in his ability to alter the news "to favor whoever he thinks will serve his business interests" (Krugman 2007). Recently, however, Murdoch has divulged that he wants Sky News to begin promoting the right-wing politics of Fox News Channel, which he also owns (Owen 2007).

## Contemporary Television News: CNN and Fox News Channel

Contemporary television news has become a complicated genre because it attempts to balance serious journalism and entertaining television. To a certain extent, that complication bonds television journalism to print journalism. The newspaper business must concern itself with keeping circulation up, staying within budget, and attracting advertisers in addition to its primary function of getting the news out. Television news programs must also attract advertisers and viewers in order to remain in business, a concern of producers of dramas, sitcoms, reality shows, and every other program on commercial television. How programs manage this feat depends on how they differentiate themselves from other television fare. In cable news—a form distinct from broadcast news—CNN, Fox News Channel, and MSNBC are the three major networks. The top two—CNN and Fox News Channel—maintain distinct brands to attract viewers in the crowded cable landscape.

Megan Mullen explains that the marketability of cable television significantly increased with the addition of 24-hour news programming. Time Warner, which owns CNN, HBO, TNT, and many other cable channels, is also the second largest operator of cable TV systems; it is second only to Comcast and claims 13.3 million subscribers (National Cable & Telecommunications Association 2008). Cable channels offer media corporations platforms on which to broadcast their libraries and venues for original programming. Compared to original series (both scripted programs and "reality" programming), news is cheap to produce, and the program supply is "virtually limitless," though it is often repetitive. News programming is therefore a mainstay of cable, with such channels as CNN Headline News, CNBC, Fox Business Network, The Weather

Channel, and Bloomberg News supplying news on a 24-hour basis (Mullen 2003, 133, 143).

Of the three major cable news networks, CNN and Fox News Channel are the most successful and most thoroughly branded networks. Ted Turner founded Atlanta-based CNN in June 1980, raising the ire of the broadcast networks (Collins 2004). In 1981, ABC and Westinghouse attempted to compete with Turner when they launched the Satellite News Channel (SNC), a 24-hour news network. The network suffered great losses, and Turner eventually bought it in 1983. In the early 1980s, then, two 24-hour news paradigms existed. Margaret Morse notes that CNN offered "rotating packages" of news, while SNC offered "constant updating and ever-changing stories." She argues that SNC's model of constantly changing news failed because it "provoke[d] anxiety" and was better suited to the "news junkie." CNN's repetitiveness, by contrast, was more comforting (Morse 1986, 73).

CNN's saving graces were the "one or two turn-on-your-TV stories" per year, which Turner used to secure greater advertising revenue (Collins 2004, 42). The network boasted $291 million in revenue and 54 million subscribers by the end of 1989 (43). After a significant ratings boost after the 1991 Gulf War, CNN's ratings dropped and continued to plummet dramatically. From 1993 to 1994, ratings dropped over 20 percent (53). CNN's "breaking news" formula was proving to be a failure, but prolonged coverage of the O.J. Simpson story in the mid-1990s reversed CNN's fortunes. In 1996, Time Warner bought Turner Broadcasting System (TBS) and CNN became part of a media conglomerate; as a result, 700 employees of TBS were fired (64). When Time Warner merged with AOL in 2001, 400 CNN employees also lost their jobs (164).

In addition to job cuts, in 2001, CNN received a superficial makeover. WB Network founder Jamie Kellner, who was appointed head of TBS after the merger, noted that CNN had no promotion system in place and was using dated visuals. Kellner instituted stylistic changes: CNN's studio became more brightly lit, its on-air personalities began to look more stylish, and its sibling, Headline News, became bloated with graphics. Hollywood visuals and entertainment values had found their way to CNN (Collins 2004, 165–166).

In 1995, Rupert Murdoch announced the beginnings of Fox News, a "really objective news channel" to balance the so-called liberal slant of CNN (ibid., 66). Also in 1995, Microsoft and NBC announced the launch of MSNBC, a cable news network that targeted CNN. Murdoch hired Roger Ailes—media adviser to former presidents Richard M. Nixon and Ronald Reagan and to George H. W. Bush when he was vice-president—to head Fox News Channel. Together, Murdoch and Ailes created a television news channel that catered to a talk-radio audience and offered a conservative alternative to the so-called liberal media (Project for Excellence in Journalism 2005, "Economics" section). At first, Murdoch had to pay cable operators to carry Fox News Channel, but within a year, Time Warner Cable began to carry Fox News Channel in New York City. Five years after it launched, the network had become a huge success; in 2001, Fox News Channel posted net revenue of $208 million. Motivated by Fox News Channel's success, MSNBC followed Murdoch's network into "opinion journalism," dropping *The News with Brian Williamson* for *Donahue*. CNN executives decided that the network's best branding strategy was to remain "basically newsier" than its two main cable rivals (Collins 2004, 194).

Cable news continues to develop and influence journalism on television. CNN and Fox News Channel, in particular, are singled out for their impact on television news and on the decisions of voters and policymakers. CNN's use of satellite technology and reliance upon breaking news transformed the news landscape, perhaps transforming politics as well. The "CNN effect" once described CNN's position as the central source of information to everyone involved in the 1991 Persian Gulf War, but the term has come to refer to the process by which the way the network frames news stories actually influences government policy (Robinson 2002, 2). According to Susan Carruthers, in the "most extreme form" of the CNN effect, "graphic television images from a crisis zone" force policymakers to shuffle priorities to address the disaster that is most visually prevalent in the news (2000, 207). It is important to note that although the term explicitly names CNN, the effect transcends any one news network. In his analysis of the U.S. government's responses to crises in northern Iraq, Somalia, Bosnia, and Kosovo, Piers Robinson posits that the CNN effect is by no means a monolithic phenomenon; he finds that it can manifest with strength or weakness (2002, 37–41).

Furthermore, the effect works in more than one way. Steven Livingston describes the CNN effect as an interaction between the news and the government: "A good way of thinking about the CNN effect is to think about the relationship between government officials and the media as a sort of dance, and . . . at various points in time it's the media who are leading in this dance," but in other instances, "government officials are using the news media to move their agenda items further up the pay scale" (quoted in Eagleburger et al. 2003, 67–68).

Fox News Channel has also been highly influential in U.S. political culture and in the television news industry. The "Fox effect" refers to the way that Fox News Channel has affected the branding strategies of the other news networks and the attitudes of voters. Fox News Channel's ratings success initiated a programming overhaul on MSNBC. Although executives said otherwise, CNN appeared to follow Fox News Channel's lead as well. CNN's leadership set up a meeting with Republicans "to discuss CNN's perceived liberal bias," and CNN began to mimic Fox News Channel's symbolic gestures of patriotism by displaying a graphic of the U.S. flag (Rutenberg 2003). CNN's flag eventually came down, but MSNBC retained theirs and implemented other strategies in an effort to improve ratings. MSNBC even fired its liberal host, Phil Donahue, and hired two conservative commentators, Joe Scarborough and Michael Savage, to "add political equilibrium to its lineup" (ibid.). While CNN eventually returned to its ostensibly neutral and international brand, MSNBC, which had no coherent brand, gravitated toward Fox News Channel's brand.

Stefano Dellavigna and Ethan Kaplan attach another layer of meaning to the term "Fox effect." In their study of voting behavior in the 1996 and 2000 presidential elections, Dellavigna and Kaplan found that Fox News Channel had a "significant effect" on voters in cities and towns with Fox News Channel. Fox News Channel "convinced 3 to 28 percent of its viewers to vote Republican," which ultimately means that "the Fox News effect could be a temporary learning effect for rational voters, or a permanent effect for nonrational voters subject to persuasion" (Dellavigna and Kaplan 2007). The Fox effect rests on conservative rhetoric that induces mimicry and apparently wins votes, while the CNN effect depends on a complex relationship between government leaders and their television sets.

In size and brand, CNN and Fox News Channel are opposites. Of all the news channels, CNN is the largest, employing 4,000 people in thirty-six news bureaus, twenty-six of which are international. In 2007, $287 million of its $687 million budget went to programming. Fox News Channel trails CNN in both size and budget, supporting 1,200 employees in fourteen bureaus, three of which are international. It spent $319 million of its $487 million budget on programming in 2007—up 20 percent from the $266 million it spent in 2006. Although both networks have spent more money on programming from 2006 to 2008, the number of bureaus for each network has dropped since 2005 (Project for Excellence in Journalism 2008, "News Investment" section). CNN has shed three of its bureaus (two of them overseas) since 2005, while Fox News Channel also dropped three bureaus (all three of its overseas bureaus).

Since the 1991 Persian Gulf War, CNN has built and attempted to maintain a brand name associated with objectivity, reliability, and global reach, but the liberal-conservative dichotomy that Murdoch and Ailes promote threatens the strength of CNN's brand association with balanced coverage by labeling it a "liberal" network (Project for Excellence in Journalism 2005, "Economics" section). The notion of an oppositional relationship between CNN and FNC extends to the territory of ratings, as well, where the two networks' importance in the industry owes in part to their relatively large audiences. Before the 2003 invasion of Iraq, CNN's ratings peaked during the month of the World Trade Center attacks. In September 2001, CNN had an audience of 2.1 million viewers, which then dropped to below 1 million viewers for twelve of the seventeen months before March 2003. Fox News Channel, however, peaked in October 2001 with 1.5 million viewers and remained above the 1 million mark for all of the subsequent months until March 2003. After the 2003 invasion, Fox News Channel maintained an audience of over 1 million for the rest of 2004, climbing to over 2 million in the two months preceding the November 2004 presidential election. CNN did not manage to exceed the 1 million mark until just one month before the 2004 presidential election (Project for Excellence in Journalism 2005, "Audience" section). In spite of CNN's superior size and less polemical brand, Fox News Channel's audience size has clearly surpassed that of CNN.

Fox News Channel has the largest median prime-time audience among cable news networks, boasting 1.4 million viewers in 2007 compared to CNN's 736,000. MSNBC's median prime-time audience is approximately 490,000 viewers. However, CNN continues to have the largest cumulative audience, the number of distinct viewers who watch over a fixed period of time (as opposed to ratings, which measures how many viewers are watching at a given moment) (Project for Excellence in Journalism 2008, "Audience" section). In 2004, Fox News Channel had twice the prime-time viewership of CNN. In the first half of 2008, Fox News Channel led CNN by only 20,000 viewers. Change was significant in 2008, when CNN attracted 170,000 more nightly viewers than it did in 2004, MSNBC drew 181,000 more, and Fox News Channel lost 90,000. Fox News Channel retained its title of "the most-watched cable news channel overall," but CNN and MSNBC experienced more growth (Steinberg 2008). MSNBC was still in third place in 2008, but its rebranding and counterprogramming strategies resulted in a modest but important ratings win in June 2008 when *Countdown with Keith Olbermann* beat Fox's *The O'Reilly Factor* by 5,000 viewers during the Democratic Party primary (Thielman 2008).

Fox News Channel entered the cable news industry as a challenge to CNN's dominance and perceived political slant. Negotiations between the two conglomerates about ownership and programming delivery systems support the perception of frequent conflict between CNN and Fox News Channel, and Turner and Murdoch are rumored to clash (Project for Excellence in Journalism 2005, "Audience" section; Proffitt 2004, 8). Whether CNN and Fox News Channel actually are polar opposites politically and stylistically is of little consequence on one level since each network's brand has thoroughly differentiated it from the other. In terms of style and content, however, the similarities between the two networks that I found attest to a lack of diversity in television news, a homogeneity that predated the contemporary cable news landscape. Edward Said wrote of the crisis in the supply of U.S. news in the early 1980s: "Together, the small group of principal news suppliers and the extraordinary array of much smaller suppliers that are independent of and yet in many ways dependent on the giants furnish an *American* image of reality that does have a recognizable coherence" (1997, 55; italics in the original). The number of mainstream and independent news suppliers has risen, but the U.S. "image of reality" remains. As Said wrote, "Despite the extraordinary variety [in media], there is a qualitative and quantitative tendency to favor certain views and certain representations of reality over others" (49).

## Conclusion

CNN and Fox News Channel are participants in a rivalry discursively based on opposing political viewpoints and materially based on ratings. Politics and economics commingle in the reputations and branding of the two networks. The rightward shift in the tone of news may have come about because of Reaganite ideology, Reagan's pro-business policies, or simply the need to reach the most affluent audience in the 1980s. Regardless of the source of the shift, the Reagan era changed the way television news is produced and provided the lax regulatory framework in which high concept flourished.

That television news and high concept converged eventually is a testament to the desire of corporate capitalists to sustain the systems that support them. Television news has transformed from a public service into a profit center that operates more or less overtly in the service of corporate parents. The idea of public service lost the tenuous grip it had on television when the networks began expecting their news programs to turn a profit in the 1980s. By incorporating entertaining formats, graphics, and narratives into their programming and by decontextualizing and simplifying events, news producers have proven their commercial value to their media conglomerate owners and to the government. In his 2003 preface to the 25th-anniversary edition of *Orientalism,* Edward Said lamented CNN and Fox News Channel's habit of "recycling the same unverifiable fictions and vast generalizations so as to stir up 'America' against the foreign devil" (Said 2003, xx). The recycling, the fiction, the ratings, the profits, and even the cheerleading call for a nuanced look at the confluence of high concept and television news.

High concept nurtures itself by promoting a lifestyle that, according to Wyatt, ignores controversial politics and promotes consumerism. Correspondingly, high-concept coverage of the 2003 invasion of Iraq avoided the depth of controversy the war incited and focused instead on appealing to the audience through a number of strategies. One of the most significant strategies was the retelling of the official narrative the government used to justify war.

# 2

## *The High-Concept War Narrative*

On March 22, 2003, an incident in Kuwait exposed the practice of speculation on television news and the effect that process had on the crafting of a narrative. That day an unidentified assailant threw hand grenades into the tents of some commanding officers of the 101st Airborne at Camp Pennsylvania in Kuwait. The prevailing belief on CNN and Fox News Channel was that the perpetrators of the crime were Arab terrorists. After several hours of repetitive conjecture about how those terrorists had infiltrated the heavily guarded camp, the U.S. military confirmed that the suspect in custody was not an Arab terrorist at all. He was a U.S. soldier. The reality of unfolding events conflicted with the simple story CNN and Fox News Channel had shaped, illustrating how journalists can narrativize events in a compelling way.

The narrativization of events on television news invites a set of complications that are distinct from the fundamentals of high-concept films. The high-concept film, a multifaceted commercial package, needs a simple story for the rest of the mechanism to fall into place. The high-concept narrative is not the stuff of high art; on the contrary, the story's simplicity is supreme, and it manifests in several ways. The single-sentence pitch is one of them, and its brevity is not just a cliché. Justin Wyatt notes that in a sample pitch for the 1990 film *Days of Thunder* ("*Top Gun* in race cars"), "one can see the movement of the narrative from the single-sentence concept" (1994, 17). In the pre-sold product of *Top Gun* and its recontextualization into the world of motorsports, we can glean the "establishment, animation, intensification, and resolution of the plot structure, as well as the star, the style, and genre of the film" (ibid.). A description this succinct would do a disservice to a more thematically complicated film like *Terms of Endearment*, Wyatt notes. In high concept, the story's simplicity is necessary in order to create the stylized and easily consumed marketing campaign. The film's lack of depth is part of the high-concept design.

That same absence of depth was evident in the reporting of the grenade attack. By repeatedly suggesting that Arabs had attempted to murder U.S. military officers in their tents, CNN and Fox News Channel journalists adhered to and manufactured a simple yet enduring story. But just as a single-sentence description cannot do justice to a multilayered film, a reductive news story cannot adequately explain complex issues like an act of mutiny or an act of war. As Robin Wood (1986) notes in his discussion of Reaganite cinema, this type of reductionism plays a role in manufacturing a reactionary worldview that is ultimately regressive. Daniel Hallin (1986a, 1986b), Robert Stam (1983), and Brian McNair (1998) argue that ideology is always present in journalism. An examination of high-concept news demands attentiveness to the ideological assumptions and political implications of the narrative the news constructs.

This chapter will examine the narrative for the 2003 war that was constructed by the state and transmitted by CNN and Fox News Channel. Central to the narrative is the marketable concept. The marketable concept of the 2003 invasion stated that the United States used its superior technology to exact vengeance for the attacks of September 11, 2001, and that its actions were morally justified. (Embedded in this simplistic narrative are the assumptions that Saddam Hussein was connected to Al-Qaeda and was harboring weapons of mass destruction.) The narrative that CNN and Fox News Channel personnel constructed to articulate that marketable concept within the first days of the war followed this simplistic storyline. This chapter will examine the practice of narrative in television news, specifically in six incidents that took place during the first days of the 2003 invasion: the opening-night "decapitation" strikes; the "shock and awe" bombing campaign; the Iraqi celebrations at Safwan; the flag-raising incident at Umm Qasr; the discovery of an alleged chemical weapons facility at Al-Najaf; and the capture of U.S. POWs. Let us first turn to the subject of narrative in television news.

## The Narrativization of News

Television news in the United States tends to transform events into stories with causes and effects. The combination of visuals and narration in television news underscores its reliance upon the narrative language of audiovisual media. Whether one points to film or literature as the cornerstone of television news, scholars in various disciplines disagree about how the news narrativizes events (Hallin 1986b; Johnson-Cartee 2005; Schudson 1995; Sperry 1981; Stam 1983; Wittebols 2004; Woodward 1997) and if the news narrativizes events at all (Dunn 2003; Ytreberg 2001).

At the core of arguments that journalists do not use narrative in their delivery of news lies notions of confirmable evidence and journalistic objectivity. Espen Ytreberg does not accept the "central premise . . . that television news has been conquered by narrative" (2001, 357). He considers that assumption to be flawed and shortsighted, and he offers an alternative explanation for what others have dubbed the narrativization of news. Using text-type theory to examine Norwegian television news in the 1980s and in the 1990s after deregulation, Ytreberg argues that the coherence of news stories is not an example of narrative but "a composite mode of description" (360). He writes, "Descriptions do not elaborate on the uncertain; they are geared towards rendering the properties of objects, actions, and happenings. . . . Description is typically not concerned with the possible or the plausible (verisimilitude), but with the strictly verifiable, that which may be given as unequivocal evidence" (ibid.). Ytreberg's analysis, though compelling, does not apply readily to war coverage that takes place over a long period of time. He concedes that a reporter's access to a "total news event" could mean that a narrative approach becomes the dominant mode of news coverage; this is what happened during the nonstop war coverage in late March of 2003 (365).

Anne Dunn applies Ytreberg's findings to radio news, and she, too, is reluctant to accept that narrative is part of television news. Like Ytreberg, Dunn cites the inverted pyramid—the fundamental model for writing news stories in which the journalist places the most important information at the beginning of the story and the least important information at the end—as proof of the absence of narrative in news. She points out that the model is informational rather than narrative in nature and that its creation is linked to the ideal of journalistic objectivity (2003, 114). News stories that follow the structure of the inverted pyramid do not fit the structure of classical Hollywood narrative, according to Dunn. She concedes that some "soft" news stories use narrative strategies, but she does not agree that this practice creates a "connected sequence of events, linked together 'through a process of transformation'" (Edward Branigan quoted in Dunn 2003, 115). Dunn allows for the possibility of narrative in news, but only when "a connection is constructed between stories . . . provided some consideration of motives or intentions is part of the connection made" (2003, 116). In that case, the episodic and compartmentalized style of news stories can succumb to the enforced construction of an overarching narrative.

Although a narrativized connected string of events might not be plausible in a traditional news cycle, the persistent barrage of 24-hour news (and war coverage, specifically) offers a radically different paradigm—one in which it is possible to see constructed stories with events linked by cause and effect, obstacles, motivations, disequilibrium, and equilibrium in a concentrated time frame. David Bordwell's explication of film narrative supports the idea of the construction of news narratives. He draws from the work of the Russian formalists to construct a theory of narration that relies heavily on the terms "fabula" and "syuzhet." Fabula refers to a series of events that can be linked to cause and effect, and syuzhet refers to how those events are arranged and presented in a film (1985, 49–50). Bordwell's definition of narration is "the process whereby the film's syuzhet and style interact in the course of cueing and channeling the spectator's construction of the fabula" (53). That construction relies a great deal on a viewer's perception of what causes what, which Bordwell attributes to the syuzhet, the way the filmmaker has presented the events of the fabula to the audience. If the way in which a particular story is told translates narrative events into comprehensible causes and effects, then the syuzhet—the arrangement of these events—can illustrate how the very structure of 24-hour news and the reliance on official sources creates meaning.

Twenty-four-hour coverage of the 2003 invasion of Iraq had two structural constraints that affected the transmission of information: limited resources (few journalists in Baghdad, for instance) and Department of Defense censorship (through the embedding system and general wartime reporting protocol). One of the ways that the syuzhet shapes the story (or our perception of it) is through the management of "the quantity of fabula information to which we have access" (Bordwell 1985, 54). This relates not only to the way that the Department of Defense shaped the war narrative, but also to the way that CNN's and Fox News Channel's syuzhet was structurally bound to the information the Department of Defense controlled and offered. The networks' syuzhet, was constrained by information "retardation" through delay or omission (56). For instance, gaps in the flow of information—as in the case of lack of knowledge about the status of Saddam Hussein—could hardly have been helped, but they created anticipation, as "temporary gaps" (in Bordwell's assessment) are intended to do (55). And as I will illustrate shortly, in some cases Fox News Channel's syuzhet delayed the transmission of information even after the Department of Defense had made new announcements.

Another consequence of the 24-hour format is repetition, and the dearth of information available to the networks at the start of the war exacerbated this tendency. The on-air personalities' repetitiveness paralleled the "redundancy" function of the syuzhet. In filmic narration, redundancy is functional and significant and works to "reinforce assumptions, inferences, and hypotheses about story information" (Bordwell 1985, 56). Redundancy in the coverage of the war was a result of the structural limitations noted above. However, I will show that the networks' repetition served narrative aims. CNN and Fox News Channel could verify few events beyond a doubt in the 2003 war coverage, so the networks constructed the war narrative based on speculation and analysis that elaborated on the uncertain. In other words, the syuzhet employed by CNN and Fox News Channel not only promoted causal connections between the events of the 2003 invasion, it also advanced hypotheses about motives and intentions.

Film theorists like Bordwell did not intend for their work to be translated to television news, but other media scholars explicitly address the potential for crossover. Robert Stam posits that television news has inherited its discursive mode of operation from both cinema and journalism, noting that "filmic procedures" and cinematic continuity are fundamental to the process of creating television news (1983, 33–34). Stam naturalizes the link between fiction and news to help explain why news is a "pleasurable" experience for viewers. The narrative form is recognizable and understandable, he argues, so the use of that form ensures that news programming will be comfortable and pleasing for viewers. Stam quotes Reuven Frank, former president of the NBC News Division, as saying, "Every news story should, without any sacrifice of probity or responsibility, display the attributes of fiction, of drama." Stam argues that the "nature" of news is not informative but is a "construct whose procedures resemble those of fiction." He points to the use of the phrase "top stories" rather than "top facts" on television news programs as immediate proof of the "fictive nature of the news." He goes so far as to identify news as "literally . . . fiction" (31).

Other scholars argue that narrative pervades the news to varying degrees and for different reasons. Michael Schudson points to the practice of using narrative in news not as a matter of "fictional" intervention but as a matter of convention (1995, 55). For Schudson, the "structure" of news stories *is* narrative; as a result, journalists fashion their stories around the "assumptions of representation" built into narrative (68). Karen S. Johnson-Cartee includes in this narrative form "a set of previously determined narrative structures," also known as narrative frames. These frames—structural components that make up the content of news stories—help simplify and guide the narrative (2005, 159).

Gary Woodward attributes journalists' simplification of complex issues to the efficiency and entertainment value—not the politics—of narrative (1997). In a similar vein, James Wittebols argues that the narrativization of network news is the result of economic circumstances and shifts in the structure of the television industry. Wittebols links news directly to the genre of serial drama. He argues that the financial imperatives of conglomeration have fostered an increase in the soap-opera storytelling format on television—an aggressively serialized form of narrative whose ability to lure and keep loyal audiences has made it the "commodity form" of television (2). He points to such genres as wrestling, reality television, and news as evidence that the soap-opera paradigm has permeated the landscape of television in the service of audience-building and profit. Network cost-cutting and the deregulation of broadcast television in the 1980s—two events that pushed network news programming away from its public service duties—resulted in a shift toward narrativizing the news in a way that would grab the attention of viewers.

In addition to attracting audiences to news programs, the narrative format has served a larger purpose in the context of a network's programming lineup. Over time, the broadcast networks positioned news programs to maximize their commercial potential. Sharon Lynn Sperry argues that because slots for newscasts precede the networks' prime-time lineups, the news is there to attract an audience that will stay with that channel's progression into the prime-time hours. She asserts that the networks intend to give audiences the same narrative format in the news that they expect from the prime-time programs. As a result, audience-building and news reporting comingle in news programs, yielding "a blend of traditional, objective journalism and a kind of quasi-fictional prime-time story-telling which frames events in reduced terms with simple, clear-cut values" (1981, 297). The narrative format provides an engaging transition that promotes upcoming programming while imposing that programming's values on the news.

Sperry's argument about audience retention is not fully applicable to 24-hour cable news networks, which obviously do not broadcast scripted series. However, flow and audience retention are key to any commercial television enterprise. Cable news networks rely on license fees and advertisers for their revenue and, like all cable channels, brand themselves to appeal to certain audiences who will attract specific advertisers. Like high-concept films, news networks seek to distinguish themselves in an industry characterized by sameness. The brand, which is essentially each network's marketable concept, helps shape the different ways that news networks narrativize events. Each commercial news network projects its brand through promotional spots and promotional speeches by on-air personalities. The respective brands of CNN and Fox News Channel—discussed briefly in chapter 1—maintain continuity across their programs. The same news stories appear on each channel, but the difference lies in how the news personnel report those stories. That differentiation is most apparent in the narrativization of news events that take place over long stretches of time.

### Narrative and Ideology

According to McNair, ideology is embedded in journalism, just as Wyatt argues that ideology is embedded in high-concept filmmaking. The type of narrative suited to high concept—simplified and easily comprehensible—makes it compatible with a reductive Reaganite conservative ideology, in Wyatt's assessment (Wyatt 1994, 195). Dunn argues that when it is narrativized, so-called hard news becomes "depoliticiz[ed]." She cites as evidence the simplification and omission of "issues and processes" as well as the way that institutional problems are personalized (2003, 115). This, I would argue, is not depoliticization. The politics remain, albeit in altered form. Simplifying the events that constitute the news and overlaying them with a narrative structure produces a storyline in which the power structures become benign or invisible; this is an inherently political exercise.

Sperry would disagree. She argues that politics do not inform narrative and that within the larger pattern of meaning production in news, the individual news stories consist of "a recognizable plot of action which sets character against character in a struggle to redeem the world" (1981, 303). Those individual stories are linked to each other and are collectively tied to the "hero plot" that is the cornerstone of so much television programming. The hero plot consists of a conflict that must be resolved by a protagonist acting for the good of all people against evil. Sperry argues that news programming's attachment to this plot may be an affirmative response to the question, "Is the world safe, and am I secure?" (301). Yet, she writes, "ideology has no relevance here at all. If you tell the news as a story, and if the story form you have chosen is a heroic tale, then there must be a protagonist and an antagonist. It is not political favoritism but simply a formulaic understanding of how the world operates" (303). I would respond by asking how such a "formulaic" way of explaining the world lies outside the realm of ideology.

No form of communication, whether it is narrative, informative, or both, exists outside of struggles for power; neutrality, like journalistic objectivity, is a well-intentioned goal that is difficult to attain. In the early 1930s, Antonio Gramsci wrote that "private initiatives and activities . . . form the apparatus of the political and cultural hegemony of the ruling classes" (1997, 258). McNair concurs, labeling owners of media the "key players" in the global arena of capitalism. He continues:

> Understandably, they use their media to support [capitalist] economies, and to preserve socio-political systems which allow them to go on generating profits. They are part of the supporting ideological apparatus of capitalism, using their media to reproduce and reinforce the values of free enterprise, profit and the market. The cultural power of the media, such as it is, is harnessed to the maintenance of certain ideological and political conditions, from which the economic health of the media enterprise derives. Culture, ideology, politics, and economics are linked in the output of media organizations in a way that is true for no other sector of capitalistic enterprise. (McNair 1998, 103)

McNair invokes Louis Althusser's "ideological state apparatus" to describe nontraditional sites of repression (like media) that serve the interests of the state.

The idea that the commercial news media's "formulaic understanding" of how power manifests and functions is somehow devoid of ideology is a dangerous one. Stam illustrates this with his exploration of the fictive "we" in television news. News programming's on-air personalities consistently corral viewers into complicity with or support of the events on the news by using the pronoun "we." The fictive "we" works to support binary oppositions and one-dimensional character types because, as Stam writes, "'We' can then speak warmly about 'Ourselves' and coldly about whoever is posited as 'Them'" (1983, 39). For Edward Said, the fictive "we" is omnipresent in the media's descriptions of "the way in which overtly Muslim countries . . . threaten 'us' and our way of life" (Said 1997, xi). Stam warns of "profound political consequences" from that sort of rhetorically enforced collusion:

> Television news, then, claims to speak for us, and often does, but just as often it deprives us of the right to speak by deluding us into thinking that its discourse is our own. Often it gives us the illusion of social harmony, the ersatz communication of a global village which is overwhelmingly white, male and corporate. (1983, 39)

Stam thus implicates television news in perpetuating the interests of specific dominant groups.

The ideological role of television news is evident even beyond the level of rhetoric. Hallin argues that aside from its functions as informer and entertainer, television news is an "ideological medium" that assembles and disseminates "packages for consciousness," or "frameworks for interpreting and cues for reacting to social and political reality" (1986b, 13). He points to the unified and simplistic news stories that prevail even in the face of "complexity and contradiction" (ibid.). Hallin sees television news as a circle; the theme begins, develops with "twists and turns dictated by problems of balance or ideological tensions the journalist may consciously or unconsciously be trying to resolve," and finally arrives at closure (28–29). He argues that television news combines narrative and ideology to support the medium's "strong belief in order, consensus, moderation, leadership, and the basic soundness of American institutions and benevolence of American world leadership," noting that the "We're Number One" jingoism of the Reagan era pushed news to the right of the political spectrum (29, 35).

It is clear that ideology informs coverage of war in the United States. Said writes,

> News does not just happen, pictures and ideas do not merely spring from reality into our eyes and minds, truth is not directly available, we do not have unrestrained variety at our disposal. For like all modes of communication, television, radio, and newspapers observe certain rules and conventions to get things across intelligibly, and it is these, often more than the reality being conveyed, that shape the material delivered by the media. . . . It ought to go without saying that the media are profit-seeking corporations and therefore . . . have an interest in promoting some images of reality rather than others. They do so within a *political* context made active and effective by an unconscious ideology, which the media disseminate without serious reservations or opposition. (1997, 49; Said's italics)

The narrativization of news is undoubtedly political. The form of news narrative differs from that of narratives in literature and film, but it is narrative nonetheless. While those like Ytreberg and Dunn, who write from the field of journalism, are hesitant to say television news employs narrative, media scholars like Stam, Sperry, Wittebols, and Woodward agree that those who report the news use narrative to lure the audience and entice them to stay. Stam, Hallin, and Said see the ideological dangers of this practice, as the form and discourse of narrative-driven news reduce complicated events to simple stories with unified themes in spite of obvious contradictions. Finally, McNair argues that ideology is important in the larger context of the global ambitions of media owners. Although scholars offer different explanations for how or why television networks use narrative as a way to structure the news, the general consensus is that in the United States, narrativization of television network news is a common practice. In what follows, I problematize that practice in the context of coverage of the 2003 war.

## The War Narrative on CNN and
## Fox News Channel

The narrativization of news in CNN's and Fox News Channel's 2003 war coverage is partially attributable to the format of 24-hour news. The format of round-the-clock news allows for prolonged coverage, discussion, and analysis of an event. Both networks suspended advertising for the first forty-eight hours of the invasion, which highlighted the focus on war. Within this context, a single narrative, which originated to a large degree with the Department of Defense, emanated from CNN and Fox News Channel. To an extent, this is not surprising; both networks had already demonstrated their reliance upon the administration's narrative before the invasion began (Dorman 2006, 14). During the invasion, the networks' personnel openly acknowledged an ongoing narrative, but the narrative discourse on each network differed. Furthermore, the networks' faithfulness to the Department of Defense's version of events was not always constant. Nevertheless, the Bush administration and particularly the Department of Defense offered a foundation upon which CNN and Fox News Channel based their marketable concept and overall war narrative.

### *The Government's Story*

The narrative of the Bush administration and the Department of Defense was essentially a tale of revenge: terrorists attacked the United States on September 11, 2001, which led President Bush to declare a war on terror and pursue Osama bin Laden and Al-Qaeda in Afghanistan. Intelligence reports provided evidence that the Hussein regime was a credible threat to the United States, and both Bush and Secretary of State Colin Powell claimed that Iraq had a significant storehouse of WMDs. The three rationales the Bush administration gave for invading Iraq—taking out Hussein, disarming Iraq, and liberating the Iraqi people—supported the broader official rationale: protecting the United States from another terrorist attack.

Secretary of Defense Donald Rumsfeld and Assistant Secretary of Defense for Public Affairs Victoria Clarke were primarily responsible for communicating the U.S. government's official position on the 2003 invasion of Iraq. In interviews and Department of Defense news briefings in March 2003, other Pentagon officials such as Deputy Defense Secretary Paul Wolfowitz and General Richard B. Myers informed the press about the objectives of the U.S. military in Iraq. For instance, in a March 12 interview with *Newsweek,* Wolfowitz declared that the intentions of the United States were not imperialistic: "We're trying very hard to make sure that the message inside Iraq is 'we're not coming with an exile government to impose some American-selected leadership on you'" (U.S. Department of Defense 2003b). He also reiterated the issue of protecting the American public and connected to that the administration's claim that Saddam Hussein posed a threat. Referring to European nations who strongly opposed the invasion, Wolfowitz said, "They're not threatened directly the way we are. They didn't experience September 11th. They're not the target of Saddam's threats the way we are" (ibid.). In his statement to *Newsweek,* Wolfowitz echoed Rumsfeld's characterization of Hussein as a "threat" in a March 11 Department of Defense briefing (U.S. Department of Defense 2003a). This message persisted across Department of Defense briefings.

Department of Defense officials also repeated specific goals in Iraq in order to keep briefings within well-defined parameters. In a March 21 Department of Defense briefing, Rumsfeld listed three overall goals and eight specific objectives. The goals were "to defend the American people, and to eliminate Iraq's weapons of mass destruction, and to liberate the Iraqi people" (U.S. Department of Defense 2003d). The objectives were to depose the Hussein regime by implementing a strategically sound and impressive campaign; to obliterate WMDs as well as their production and distribution facilities; to eliminate the terrorist threat in Iraq; to uncover terrorist networks throughout the Middle East; to uncover the international trade in WMDs; to aid the Iraqi civilians through humanitarian efforts; to "secure" Iraq's natural resources for the Iraqi people; and to facilitate Iraq's transition to a non-threatening and democratic system of government (ibid.). While the Department of Defense advanced several goals of the invasion, implying that victory was contingent upon their fulfillment, White House spokesman Ari Fleischer advanced one particular goal—disarmament—that, if met, would signal "success" (White House 2003a). Military might, WMDs, terrorism, oil, and the well-being of the Iraqi citizenry were mainstays of Department of Defense and White House topics, and CNN and Fox News Channel consistently returned to these themes in their war narratives. They provided the basis for a coherent marketable concept at both networks that emphasized simplistic ideas and avoided complexity.

## The Networks' Story

The war narrative, as constructed on CNN and Fox News Channel, had a distinct beginning and ending. The two networks placed the beginning of combat on March 19 and they supported the Bush administration's claim that combat ended on May 1, when President Bush landed on the USS *Abraham Lincoln* in a Navy S-3B Viking and declared that "major combat operations in Iraq have ended" (CNN.com International 2003). CNN reiterated that closure date on its special "War in Iraq" Web pages; it inserted this note at the top of the main page: "This page was archived in May 2003 when President Bush declared an end to major combat" (CNN.com 2003a). Bush's declaration provided a clear end to a narrative that climaxed on April 9 when U.S. soldiers rolled into Baghdad and pulled down a statue of Saddam Hussein in Al-Firdos Square. The president's declaration that combat had ended—which of course was entirely false—supported the narrative structure that CNN and Fox News Channel had adopted.

While CNN and Fox News Channel news personnel attempted to deliver war news as though it were objective reality, they simultaneously and explicitly referred to the war *as a narrative* that had a well-defined trajectory and a predetermined ending—an assumption that echoed the message of the Department of Defense. One of the most prevalent descriptions of events linked the war to a specific kind of narrative. CNN and Fox News Channel depicted various occurrences as "dramas," "dramatic" and "unfolding dramas," and Fox News Channel's Greg Palkot even referred to a specific form of drama with his description of a "soap opera in the center of Baghdad" (CNN March 19 and 21; Fox News Channel March 20–23).[1] Taking these narratively loaded terms even further, news anchors and commentators referred to events as "plotlines" and "storyline[s]" with an imminent "second act" or an upcoming "part two" (Fox News Channel March 20–23). References to a script also sustained the idea of the war as narrative. Wolf Blitzer stressed that the last-minute alteration to the war plan was "totally unscripted" (CNN March 21). Fox News Channel's E. D. Hill said that the Iraqis were not "going by the script they thought they were going to be reading," although a colleague remarked that the war was "going as scripted" (Fox News Channel March 21). This choice of rhetoric lent credence to the idea that a speedy U.S. victory based on the goals and objectives outlined by Rumsfeld was predetermined.

On both CNN and Fox News Channel, a U.S. victory was certain and well publicized. The confidence displayed on the networks reflected that of Pentagon officials and reinforced the marketable concept. For example, General Myers asserted in the March 21 Department of Defense briefing that the United States

---

1. All of the citations in this and other chapters to coverage of the Iraq war in March and April are for the year 2003.

would prevail. He remarked, "Rest assured, the outcome is not in doubt. We will disarm the Iraqi regime and ensure their weapons of mass destruction will not fall into the hands of terrorists" (U.S. Department of Defense 2003d). The next day's Department of Defense news briefing included more of the same; Major General Stanley A. McChrystal, the Joint Chiefs of Staff's vice-director for operations, conceded that the "progress and timing" of the invasion were unpredictable but asserted that the "outcome [was] not" (U.S. Department of Defense 2003e). The predetermined outcome communicated by the White House and the Department of Defense reinforced the marketable concept by emphasizing the superior skills and technology of the U.S. military.

Journalists and news analysts at CNN and Fox News Channel seamlessly adopted the official storyline into the narrative of their own reporting. On March 20, White House correspondent John King reported that Bush had been informed that operations in Iraq would "wrap up pretty quickly now" (CNN March 20). Jamie McIntyre also reported that the Pentagon wanted to create a feeling of inevitability that the United States would win easily (CNN March 20). Retired army major general Robert Scales repeated that idea, emphasizing the value of psychological warfare: "If we can continue that whole image of unstoppable force . . . we can achieve our end" (Fox News Channel March 21). CNN anchor Aaron Brown assumed that Hussein knew "how this plays out," implying that Iraqi defeat as defined by the official goals of the invasion was assured (CNN March 20). CNN guest and former secretary of state Lawrence Eagleburger remarked, "This is the end. They know they're going to lose" (CNN March 21). Retired general Wesley Clark echoed that level of confidence when he said, "When you own the skies, and you have precision weapons . . . it's just a matter of time" (ibid.). Aaron Brown and Major Garrett both claimed that the "ultimate outcome" was undeniable, a remark that echoed Rumsfeld's statement that the "outcome . . . is determined, it's certain" (CNN March 23). Military analyst David Christian was sure that the "shock and awe" bombing campaign would "make everybody give up," and retired lieutenant general Thomas McInerney foresaw that the war would be "short" and "decisive" and would "not go a week" (Fox News Channel March 20 and 21). This statement departed from the Pentagon's warning that predictions about duration were inappropriate (U.S. Department of Defense 2003b,e). But predictions about outcome were clearly acceptable. Brian Jenkins of the Rand Corporation declared, "We will achieve. There's no question about the outcome" (Fox News Channel March 23). Greta Van Susteren wondered what would happen "after we win this war in Iraq," although she later repeated the U.S. military's promise that the "outcome is still certain" (Fox News Channel March 22 and 23).

CNN and Fox News Channel personnel noted that the Department of Defense had qualified its assurances of victory with caution about excessive optimism and anchors periodically attempted to rein in their confidence. Often this occurred in contradictory ways. For example, Sean Hannity expressed incredulity about the fact that other media outlets expected the war to be easy but then pronounced, "We're gonna win this in short order. . . . The one thing that is never in doubt, we're gonna win this and we're gonna win it sooner [rather] than later" (Fox News Channel March 24). An equally incongruous pair of statements came from Aaron Brown, who claimed that "Americans own the sky" but followed that with the warning, "No one should get the sense that this thing's a slam dunk" (CNN March 21). Despite some backpedaling, the overwhelming sense from the two networks in the first five days was that the United States (and the coalition, to some extent) would prevail absolutely. And even though no one clarified what winning the war meant precisely, the fact that CNN and Fox News Channel personnel had already crafted the idea of an ending reinforced the narrative framework they imposed upon the war.

In between the clear beginning and the ending were many incidents that CNN and Fox News Channel depicted as contributing to the overarching narrative of the war. During the first five days of the invasion, several events generated a great deal of attention on CNN and Fox News Channel, including fires in Iraqi oil wells, a grenade attack on the 101st Airborne, the Iraqi search for missing U.S. pilots, and U.S. negotiations with Turkey concerning overflight rights. In this chapter, I focus on six events: the "decapitation" strikes on March 19; the "shock and awe" bombing campaign; the Iraqi "celebrations" at Safwan; the raising of the U.S. flag at Umm Qasr; the capture of U.S. POWs; and the discovery of a suspected chemical weapons plant at Al-Najaf. The narrative continuity that CNN and Fox News Channel maintained across each complex event illuminated not only the implementation of high-concept narrative on television news but also Hallin's point that news narratives work to contain events within dominant ideological frameworks.

### "Decapitation"

On March 19, CNN and Fox News Channel personnel circulated a story initially released by the White House that pushed the official war narrative forward and provided elements of the marketable concept that the networks adopted. According to the Bush administration, CIA and Pentagon officials met with President Bush and presented him with up-to-the-minute intelligence about the whereabouts of top Iraqi officials. They told him that he needed to act quickly or he would lose an opportunity. During the four-hour meeting, Bush decided to begin the war in Iraq. The first strikes took place late on March 19 (early March 20 in Baghdad). Both CNN and Fox News Channel referred to these events as "decapitation strikes," although journalists in other news media avoided such inflammatory language, well aware that an executive order bans the United States from attempting to assassinate the leaders of another nation (CNN.com 2003b; White House 2003b).

The way that CNN and Fox News Channel represented the event served two primary functions: it publicized the stars of the invasion and it created a desire for the "shock and awe" bombing campaign. On one level, the two networks' coverage of these initial strikes laid the groundwork for a narrative with two stars, George Bush and Saddam Hussein. (Because I will fully explore the star personas the networks constructed for Bush and Hussein in chapter 3, I will discuss them only briefly here.) On March 20 and 21, General Myers noted and praised the "flexibility" of all parties in carrying out the last-minute orders (U.S. Department of Defense 2003d). Following suit, CNN and Fox News Channel commentators applauded Bush for his intelligence, action, flexibility, and leadership skills. Meanwhile, most of the commentary about Hussein dwelled on whether or not the "decapitation" strikes had taken him out. CNN and Fox News Channel on-air personalities worked earnestly to shape the narrative, placing the fate of Hussein at the forefront of discussion. Commentators on Fox News Channel characterized questions about Hussein's fate as "the great debate of this war so far" (Fox News Channel March 21). Late on March 23, Fox News Channel broke the story that an allegedly injured Hussein had been seen being dug out of his compound after these first strikes. Several days after the initial strikes, Fox News Channel was treating Bush's decision to launch the "decapitation" strikes as an action that might have inflicted damage upon the arch-villain Hussein, a move that might also have thrown his henchmen into confusion. The strikes thus provided ample opportunity for network journalists and commentators to cultivate a sense that Bush and Hussein were the stars of the wartime narrative.

The second and more important function of the coverage of the initial strikes was to provide an opening to speculate about the forthcoming "shock and awe" bombing campaign. Citing the work of Herbert Gans, Robin Andersen argues that "opening with attack and disruption is the beginning of a good story" (2006, 72). Indeed, by using one event to discuss the other, CNN and Fox News Channel pushed the narrative forward, built anticipation, and focused on the climax of the marketable concept: the ultimate display of military technology. Furthermore, discourse surrounding the "decapitation" strikes positioned the forthcoming "shock and awe" campaign as pre-sold and as a significant plot point.

From late March 19 until approximately a quarter past noon Eastern Time on March 21, the predominant issues on CNN and Fox News Channel were whether Bush had postponed the "shock and awe" bombing to take advantage of the "target of opportunity" to take out key Iraqi leaders and speculation about when the world would see "shock and awe," if it would see it at all. The amount of time network personnel devoted to speculating about when and if the bombing campaign would occur was considerable, and the networks used a wide range of strategies to build anticipation. One way involved elaborate descriptions that built tension and expectation. Wolf Blitzer first referred to "shock and awe" as "the more orchestrated, the choreographed U.S. air campaign" and later referred to it as the "massive bombardment campaign" and the "full fury of the . . . coalition" (CNN March 20). Pentagon correspondent Barbara Starr dubbed it "the massive coordinated air and ground attacks" (ibid.). For Bret Baier, "shock and awe" was the "massive bombardment" that "would be impossible to miss" (Fox News Channel March 19). Retired U.S. navy captain Chuck Nash described it as "the devastating, blinding, you'll-know-it-when-you-see-it" type of bombing (Fox News Channel March 19). The descriptions bore a great resemblance to each other, though each surpassed the previous one in hyperbole.

Other descriptions on Fox News Channel characterized "shock and awe" as "the real mass of power"; "unprecedented, never before seen in the history of warfare"; "intense explosions going all over the place, ten bombs a minute . . . for upwards of an hour"; "so overwhelming, so ferocious"; the "more dramatic wave of attacks"; "the awesome campaign"; the "big show"; the "big spectacular attack"; "a rain of bombs and cruise missiles like nothing the world has ever seen before"; "that stupendous salvo"; "hundreds and hundreds of precision-guided bombs and cruise missiles slam[ming] into the heart of the Iraqi military"; "the most ferocious aerial bombardment in history"; the "colossal explosion of ordinance"; and "the big fireworks" (Fox News Channel March 20 and 21).

Descriptors like these increased anticipation for the bombing campaign, and some network personnel made their eagerness for the high-tech display explicit. Wolf Blitzer said that "shock and awe" was what "we had all been bracing for" (CNN March 20). Other journalists incorporated the use of the "fictive 'We'" to play up the forthcoming bombing: "we had all been waiting," "everyone seems to be waiting," "the world waits," "we've of course been awaiting," "we have come to expect," and "we've all been anticipating" (Fox News Channel March 20 and 21). The extreme end of this display of eagerness came from Fox news analyst retired colonel David Hunt, who complemented the military on playing the "shock and awe" card "brilliantly, to the point where now we're actually . . . almost begging . . . 'give us some shock and awe, come on, show us some bombs'" (Fox News Channel March 20). Hunt and others highlighted the spectacle of technology and the thrill of anticipation that could only be satisfied by the unfolding of the pre-written narrative.

The atmosphere after the initial strikes constructed the impending "shock and awe" campaign as a decisive moment in the war narrative. Network journalists were preoccupied with questions about the long-awaited campaign: *if* "shock and awe" would occur, *when* it would occur, if it had *already* occurred, why it had *not* already occurred, and if the threat of "shock and awe" was simply a strategically placed piece of misinformation. Network personnel explained that the "shock and awe" campaign had been delayed by the initial "decapitation" strikes and by the Pentagon's need to assess the effectiveness of those strikes. Although they explored the possibility that the coalition would not need to employ "shock and awe" because of "massive psychological campaigns" that likely would result in mass Iraqi defections, commentators also reassured viewers that "shock and awe" was "on the agenda," "on the docket," "ready to go," "still planned," and "on the table" (Fox News Channel March 20; CNN March 20 and 21; Fox News Channel March 21). Fox News Channel personnel educated viewers about the theoretical side of "shock and awe" when Harlan Ullman, the concept's co-author, appeared on Fox News Channel on March 20 and March 21 to discuss his theory of "rapid dominance" (Fox News Channel March 21).[2] All of this talk framed "shock and awe" not just as a technological spectacle but also as a pre-sold concept and a potentially key plot point.

CNN and Fox News Channel news personnel explored the motivations for and uncertainties about "shock and awe" at great length, creating expectations, establishing an intriguing opening to the war narrative, and dutifully setting up two possible scenarios of cause-and-effect events. In the first scenario, successful "decapitation" strikes would result in mass Iraqi surrenders; in the second, ineffective "decapitation" strikes would make "shock and awe" necessary, which would also result in mass Iraqi surrenders. The theory about

---

2. The other authors of the theory of rapid dominance include James P. Wade, L. A. Edney, and the National Defense University Institute for National Strategic Studies; see Harlan K. Ullman and James P. Wade, *Shock and Awe: Achieving Rapid Dominance* (Washington, DC: National Defense University Press, 1996).

the second potential chain of events was tested when "shock and awe" began on March 21.

### *"Shock and Awe"*

The appearance of the "shock and awe" bombing campaign was significant as a spectacle, a marketable concept, and a plot point. The magnitude of the bombings as shown on television obscured all else, including the physical consequences of the explosions. CNN and Fox News Channel on-air commentators praised technology for its own sake and for the sake of the war narrative. The massive bombing campaign allowed the networks to portray a decisive large-scale act that proved the might of the U.S. military, provided a suitable action sequence for an audiovisual medium, and reinvigorated a narrative that had begun to stagnate because of the delay.

The eagerness for the bombings led CNN and Fox News Channel on-air personalities to mistakenly announce that the "shock and awe" had begun on March 20. When the bombings did begin on March 21, there was no mistake; Fox News Channel reported that the Pentagon had assured them that "A-Day," the Pentagon's term for "Air Day," was under way (Fox News Channel March 21). Network personnel uniformly described the bombings as visually spectacular and focused on the spectacle of the event instead of discussing its consequences. After the bombing, Kyra Phillips reinforced the association with made-for-television drama when she began her coverage with the words, "Welcome to shock and awe" (CNN March 21).

For the most part, CNN and Fox News Channel news personnel were enthralled by the spectacle of "shock and awe" and freely shared exuberant reactions to the bombing. On CNN, journalists characterized the campaign as "literally awesome," "just amazing," "awesome," "biblical," a "fireworks show," and "Armageddon" (CNN March 21 and 22). Fox News Channel personnel described the bombing as an "awesome campaign," an "extraordinary scene," "very impressive," "extremely awesome," an "incredible display of ordinance," "a staggering bombardment," a "stunning barrage," and "quite a show" (Fox News Channel March 21 and 22). Not all the reactions on Fox News Channel were as celebratory, though; reporter David Chater of Fox's British sibling network referred to the bombing as "a dreadful sight" and Shepard Smith once called the campaign "devastating" (Fox News Channel March 21). Brit Hume expressed disappointment that the bombing "didn't go on with the force and duration that I think that all of us had come to expect," a reflection of the way that his network had raised expectations (Fox News Channel March 21). Hume was the only Fox journalist to express disappointment about the bombing, though; Mike Emanuel felt certain that "folks watching the developments on live television were certainly not disappointed by the military might of the U.S." Harlan Ullman likewise characterized the "reaction in the U.S." as "euphoric" (Fox News Channel March 22).

After the bombing campaign, CNN and Fox News Channel personnel began to manipulate a key element of the war narrative: Iraqi citizens. Iraqi civilians were not stable character types in CNN and Fox News Channel's war narrative. Iraqi civilians were rarely visible in the images network journalists transmitted, were unable to speak for themselves during the attack, and had a propensity for complicating the official storyline Washington had written and the networks had adopted. The convergence of these three factors made them malleable figures in the narrative both networks promoted. When the mass surrenders that both networks (and the White House) had predicted as a response to the bombing campaigns did not materialize, both networks took note of the deviation from the prewritten storyline. In their analyses of the effect of the bombings, Fox News Channel journalists reported that the Iraqi military was not surrendering as expected and CNN's journalists reported that only large-scale high-level surrenders would end "shock and awe" (Fox News Channel March 21; CNN March 22). The networks' focus on surrenders put the burden for ensuring the safety of Iraqi civilians squarely on the shoulders of the Iraqi military (and not on the shoulders of the U.S. soldiers who were bombing them).

In fact Fox News Channel personnel claimed that the invading U.S. military was not a threat to Iraqi citizens. Major Garrett reiterated the Department of Defense's statement that the bombing was not about the citizens but about the regime, positioning Iraqi citizens outside the action but central to the narrative (Fox News Channel March 24). Such statements parroted the Department of Defense's recurring statements throughout the first five days of the invasion that the war targeted the regime, not Iraqi civilians (U.S. Department of Defense 2003b, 2003c, 2003d).

Fox News Channel's concerted effort to depict the Iraqis as insulated from the bombings was undermined at various points by its sibling news network, London-based Sky News. During the bombing, David Chater remarked that "the best thing" that he could do was to "try and witness this and try to give . . . some sort of impression of what it's like for the civilians who are suffering this awesome bombardment." Shepard Smith quickly cut in to temper Chater's expression of empathy by informing viewers, "Of course the idea behind this awesome bombardment . . . and the civilians who are enduring it . . . is to free them" (Fox News Channel March 21). At another point, a Sky News anchor asked his correspondent how "shock and awe" had "affected the ordinary people" in Baghdad, but before he could finish his question, Fox News Channel's Bob Sellers interrupted the feed and changed the subject (Fox News Channel March 22).

While Fox personnel took great care to maintain the image that ordinary Iraqis were safe, CNN journalists spent a token amount of time speculating about or assessing the impact of the bombing on the citizens. May Ying Welsh, reporting from Baghdad, stated that everyone there was "definitely" affected by the bombing and that contrary to Department of Defense reports, civilians did live near the government buildings under attack. Aaron Brown conceded that the attacks were "terrible . . . if you're on the other end of them," and after eight minutes of "shock and awe," Wolf Blitzer remarked that "we can only imagine what terrifying state most of those people are presumably in" (CNN March 21). These sentiments starkly contrasted with but did not destabilize the upbeat mood surrounding "shock and awe."

Immediately after the bombings, references to the Iraqi citizens in Fox's coverage fell into two categories: comments about how they would be liberated and observations that the United States had refrained from disrupting the electricity grid in Baghdad because the war was being waged against the regime and not against civilians. Fox personnel emphasized the U.S. goal "to limit collateral damage," which they interpreted as evidence of a "humane" war (Fox News Channel March 21 and 22). David Hunt characterized the U.S. position as benevolent when he said, "We could flatten this country, and we're still trying to be humane while we're conducting a war" (Fox News Channel 20 March 2003). Hunt's remark echoed Rumsfeld's claim that the invasion was a "humane effort" by the coalition (U.S. Department of Defense 2003d). Hunt's comment and the comments of other network personnel that accompanied images of "shock and awe" emphasized the technological superiority of U.S. weaponry and its direct relation to the well-being of Iraqi citizens. By describing the invasion as an act of goodwill and by repeating this description throughout the destructive bombing campaign, the networks clung to the marketable concept while forging the type of nationalist consensus that Edward Said discusses. He writes, "The notion that American military power might be used for malevolent purposes is relatively impossible within consensus, just as the idea that America is a force for good in the world is routine and normal" (1997, 53–54). Nationalist consensus sets boundaries around the media's discussion of U.S. force, and in 2003, these boundaries effectively excluded discussion of what was happening to Iraqi civilians as the U.S. military bombed their country.

Both CNN and Fox News Channel journalists and military analysts de-voted a great deal of time to discussions of the use of precision weapons in the bombing campaign as evidence that these weapons were safe for Iraqi civilians and therefore illustrated the humanity of the war effort. This strategy also rein-forced the idea that the United States was not only exacting revenge on behalf of U.S. citizens but was also liberating Iraqis from an oppressive regime. Keeping the Iraqis safe through superior military technology was thus a crucial part of the marketable concept. Rumsfeld repeatedly assured the press that the coali-tion was taking "every precaution to protect innocent civilians" (U.S. Depart-ment of Defense 2003c). On CNN, Wesley Clark affirmed that "shock and awe" was "precision bombing . . . not carpet bombing . . . not directed at populated areas." Kyra Phillips consistently asserted that precision weaponry would cut down on the number of civilian casualties (CNN March 21). Nic Robertson, who was in Baghdad, even said that he felt safe during the bombing. Aaron Brown reiterated that the weapons were precise, but he was somewhat more realistic about the possibility of civilian casualties (CNN March 22). On Fox News Chan-nel, Shepard Smith called "shock and awe" "the most . . . precise attacks in the history of warfare," and Burton Moore spoke of how the accurate weapons minimized the risk of hitting civilians (Fox News Channel March 21).

Discussions of precision weaponry also gave on-air personalities the op-portunity to recap the war narrative by reiterating the possibility that Iraq's WMDs could result in massive casualties. For example, when Brit Hume asked retired lieutenant general Thomas McInerney about the relationship between precision-guided weapons and casualties after the "shock and awe" campaign, McInerney replied that precision weapons would ensure that there would be fewer casualties than there were in the 1991 Persian Gulf War. McInerney then steered the conversation toward WMDs and the possibility that Hussein could "unleash his own weapons" and place the blame on the United States (Fox News Channel March 21). The massive bombing campaign allowed McInerney to reinforce the superiority of U.S. technology while reminding viewers of the justifications for war.

In an exercise in projection, news personnel assumed that Iraqi civilians shared their confidence about the accuracy of U.S. weapons. Throughout their coverage of the war, on-air personnel in the United States complemented report-ers' eyewitness accounts of safe Iraqi citizens with comments about precision weapons. John Burns reported that during the "shock and awe" campaign in Baghdad, "Iraqis wandered out of their homes, hotels, on the embankment on the east side of the Tigris, and went forward to the river to get a better look. . . . These people . . . have an almost complete confidence that . . . [the U.S. military] have got the coordinates right" (CNN March 21). A more in-depth example of this sort of representation of Iraqi civilians appeared in Fox News Channel's reporting on Iraqi refugees. CNN and Fox News Channel news personnel reported that most of the refugees arriving in Jordan were African students and not Iraqi citizens. Fox News Channel reporters, in particular, introduced the theory that the absence of Iraqi refugees was a direct result of the military's use of precision weapons.

Department of Defense briefings had already established that Iraqis did not need to leave their homes during the bombings. On March 20, Rumsfeld stated, "There is no need for Iraqis to flee across their borders into neighboring countries" (U.S. Department of Defense 2003c). This statement assumed a great deal of significance for Fox News Channel personnel, who took the statement as evidence that Iraqi civilians were safe. Early on March 22, Bob Sellers asked Jennifer Eccleston, who was reporting from Jordan, if the low number of Iraqi refugees at that location was related to the Iraqis' awareness of the military targets and precision weaponry. Eccleston replied that the more logical explanation was that the road to Jordan was a "battle zone" and the trip was expensive. She also pointed out that many Iraqis had sought shelter in the countryside. A little over an hour later, Steve Harrigan, also reporting from Jordan, voiced several potential reasons for the dearth of Iraqi refugees: the Iraqis might have been afraid, they may have lacked the money to leave, and they might have interpreted "this military action as a strike against the leadership" (Fox News Channel March 22). Within just a short period of time, a reporter based in Jordan was expressing a speculative statement about the motivations of Iraqi civilians that had originated in the Fox News Channel studio with Bob Sellers.

The next refugee report came later in the day, when Todd Connor reported from Jordan that the Iraqis might have decided to remain in their homes because they anticipated regime change and knew that U.S. bombs were "accurate." Connor's subsequent report a short time later reiterated that "others" believed that the Iraqis had stayed behind because of the expected regime change. He also noted that "speculation" existed that Iraqis felt safe because of the precision weapons (Fox News Channel March 22). In a later report on refugees, studio-based news anchor Julian Phillips hypothesized that Iraqis might have felt that the war was not against them. Shortly thereafter, Colonel Patrick Lang assessed the refugee situation and agreed, saying "We've convinced the population that we're not the reincarnation . . . of the Mongols" (Fox News Channel March 23). When Steve Harrigan gave his next report on refugees, Phillips prompted him by saying, "You had a theory [about the lack of Iraqi refugees]. Could you tell our audience what your theory is once again?" Harrigan admitted the possibility of "a lot of factors," but he added that the Iraqis might have thought the war was "not against the Iraqi people" and that "their best bet" was "to ride it out" (ibid.). When Phillips asked Major Dana Dillon for his insight on the refugee situation, Dillon responded, "It's been fairly precise. . . . Their homes aren't being burned down. . . . They probably just want to stay home and wait the war out" (ibid.). Tony Snow remarked that Iraqis were going on with their daily lives because they had a "great deal of faith in American technology" (ibid.). Speculation regarding the lack of Iraqi refugees continued without any effort to verify the claims that Iraqis were safe and trusted in the U.S. military.

While the military, government, and mainstream journalists stressed the idea of a safe invasion, they also refused to count casualties. The humanity of the invasion was a ubiquitous talking point precisely because the material consequences went unrecorded by military officials. Bodies were counted, however. According to Iraq Body Count, an independent organization that uses multiple official sources to document violent deaths suffered by Iraqi civilians, 3,976 Iraqi noncombatants died during March 2003. The number was slightly lower in April 2003, when 3,437 violent deaths occurred. This inhumanity found no place on CNN or Fox News (Iraq Body Count 2009).

The "shock and awe" bombing campaign propelled the war narrative just as speculation about the delayed follow-up to the "decapitation" strikes almost stalled the narrative's progression. The bombings gave network news personnel the opportunity to reflect at great length on the superiority and precision of U.S. weapons. At both networks, journalists interpreted this as evidence that the United States was conducting a "humane" war, thus perpetuating key elements of the marketable concept: superior military technology employed by a vengeance-seeking but nonetheless human military force. In the service of this narrative, and with very little evidence to support their claims, Fox News Channel personnel repeatedly spoke about and for Iraqi citizens. Fox news personnel tended to push certain themes to the extreme, dwelling on speculation and repeating the official line of the Department of Defense as often as possible. CNN personnel followed this script as well, albeit to a lesser extent; they at least attempted to complicate issues that Fox personnel posited as clear-cut. In sum, the "shock and awe" bombing campaign solidified the themes of the narrative while providing a spectacular transition from the brief attacks at the start of the war to the faces and events of the ground war.

### "Celebrations" at Safwan

When the massive bombings began, Shepard Smith introduced the campaign with the following statement: "The shock and awe phase of the war to liberate the people of Iraq and disarm the Iraqi regime has begun." This announcement clearly linked the campaign to the overarching narrative created by the Bush administration and the Department of Defense (Fox News Channel March 21). In fact, Smith's phrasing was very similar to Rumsfeld's words in the March 21 Department of Defense briefing: "On the president's order, coalition forces began the ground war to disarm Iraq and liberate the Iraqi people yesterday" (U.S. Department of Defense 2003d). Smith reiterated his linkage between bombing and liberation at least three times that day with the exact same phrasing. Liberation was a significant part of the war narrative, and once again the networks attributed motives and attitudes to the Iraqi citizens that fit the narrative and simplified a complex situation. This simplification intensified the networks' degree of adherence to the marketable concept.

On March 21, CNN and Fox News Channel aired a video clip of a few people, presumably Iraqis, celebrating and greeting U.S. soldiers allegedly in the southern Iraqi town of Safwan. The edited video showed Iraqis smiling and shaking hands with and kissing U.S. soldiers. No more than a handful of Iraqis appeared in any one shot. Both networks used the video evidence of the "celebrations" at Safwan as proof of the liberation aspect of the war narrative, but the narrative discourse on CNN and Fox News Channel differed noticeably. Both networks welcomed Iraqi civilians into their narrative discourse much more when they had this tenuous confirmation that Iraqis were greeting U.S. troops as liberators. However, when resistance escalated and Iraqi civilians either did not receive the soldiers warmly or revealed their initial kindness to be a deception, CNN and Fox News Channel journalists adapted the evidence in a way that continued to serve the pre-determined narrative.

CNN news personnel did not interpret the Safwan video as overwhelming evidence of Iraqi support for U.S. actions, but they did not dismiss the veracity of that element of the official narrative either. On CNN, *New York Times* reporter Dexter Filkins phoned in a report from a town that had been "liberated" by the United States. He reported that "people were joyous" but "fearful" that the liberation would not last (CNN March 21). Later, Judy Woodruff presented two videos, one of "people . . . showing defiance" who were "presumably . . . pro-Saddam" and another video from Safwan of people "cheering the arrival of U.S. troops" (ibid.). Commenting on the Safwan video, Aaron Brown remarked that the Iraqis should have appeared happy to see U.S. soldiers if they had "half a brain" (ibid.). His cynical view did not deny the message of the video, but it did question its reliability. Heidi Collins did not offer much commentary in her report of the video, instead characterizing the Iraqis as simply "overjoyed" (ibid.). CNN personnel were measured in their reaction to the visual evidence, yet they still accepted the possibility that what they were seeing was evidence of liberated, joyous Iraqis.

Their counterparts on Fox News Channel had the opposite reaction. William La Jeunesse announced the footage with the statement that he had the "first video of a village that has been liberated by the U.S. troops" and said that it included "some people cheering" (Fox News Channel March 21). He elaborated, proclaiming that the Iraqis in the video were chanting, "Saddam, your days are numbered" (Fox News Channel March 21). One could read La Jeunesse's last comment as a colorful interpretation of the video, but he nevertheless was speaking for the Iraqis who appeared in footage that had not been translated into English. This same "translation" appeared on CNN (CNN March 21). La Jeunesse's tactic of speaking for Iraqi citizens mimicked the strategy of the Department of Defense; on March 20, Rumsfeld had remarked that the United States had seen "a good deal of evidence that the Iraqi people want to be liberated" without substantiating his claim (U.S. Department of Defense 2003d). Fox News Channel interpreted the Safwan video as verification of Rumsfeld's unsubstantiated claim.

Fox reports of the Safwan video were positive and frequent after La Jeunesse's initial story. Jon Scott described it as footage of a "liberated" town, and Brigitte Quinn initiated the trend of using the video as proof that the official narrative of the marketable concept was accurate with her comment that "we've already seen some scenes of celebration." La Jeunesse explicitly identified the video as evidence of liberation, remarking that the Bush administration had hoped they would see a video like that: "This is what [they] said was going to happen, that the people of Iraq would welcome the troops" (Fox News Channel March 21). He later remarked, "This is a scene we expect to see as city after city falls in the north and the east of U.S. troops." He added that the people of Safwan "are liberated, they have their freedom back, and they have American and British troops to thank for that" (ibid.). At that point in the circulation of the video, the narrative was solid, and both networks began to treat the video as evidence of a large-scale positive reception.

The next day, Adam Housely interpreted the Safwan video as a sign of continuing Iraqi rejoicing, declaring that "celebrations" were ongoing and that "a lot of people [are] happy that they have been liberated." David Asman and John Kasich added dramatic details in their commentaries about the video. Asman commented on the "incredible scene . . . where the U.S. troops were seen literally as liberators. . . . People were in tears." Kasich proclaimed that the Iraqis were saying, "Thank God you're coming here" (Fox News Channel March 22). The lone dissenting viewpoint came from Kasich's guest, Congressman Greg Meeks

(D-NY), who replied that he had only seen a continuous loop of the same shot of one person celebrating (ibid.). But by March 24, a new interpretation of Iraqi reaction to the invasion was needed as the evidence became unmistakably clear that Iraqis did not in fact welcome the U.S. invasion. Asman reported, "In Safwan, they're turning on the soldiers." As if personally betrayed, he then asked, "What is the true face of Iraqis?" (Fox News Channel March 24). La Jeunesse likewise emphasized the inconsistency of the reception to the invading soldiers when he noted that southern Iraqis waved as the U.S. soldiers entered and then offered another hand gesture as they departed.

CNN and Fox News Channel journalists welcomed the Safwan video into the news rotation with open arms. The networks had a recorded piece of history that affirmed their simple narrative: the coalition would enter Iraq, liberating grateful Iraqis along the way, and then quickly reach Baghdad to depose the regime. Disarmament was the practical objective, but liberation was the heart of the operation and a strong facet of the marketable concept.

The Safwan video represented the morality and goodwill of the United States and the coalition forces—qualities that CNN and Fox News Channel had publicized since the first night of bombing. For example, on CNN, Senator Joe Lieberman affirmed that the United States offered Iraqis "a better way, a better life" (CNN March 19). Aaron Brown likewise asserted that the children of Iraq would live better lives after the invasion (CNN March 20). Embedded reporter Walter Rodgers placed great emphasis on Captain Clay Lyle's pep talk in which he proclaimed to his troops that the United States was "invading to liberate" (ibid.). In addition to Shepard Smith's continuous references to the objective of liberation, Thomas McInerney interpreted the fact that there was electricity in Baghdad after "shock and awe" as Bush saying to the Iraqi people, "We want you to be part of that liberation" (Fox News Channel March 20). The vision of the heroic coalition liberating the disenfranchised Iraqis also permeated Shepard Smith's paternalistic statement that "the great hope of the allies" was that the Iraqis would "be able to hoist their flag yet again and live in freedom and be fed by the allies and clothed by the allies and given medicine by the allies to go on with a life that will probably be a lot more pleasing" (Fox News Channel March 21). Such statements exhibited an ironclad adherence to the marketable concept and its narrativization.

Several news personalities mentioned the idea that the United States would be seen as an occupier instead of a liberator—a baseless fear, in their estimation. These individuals repeated the theme of liberation incessantly and energetically after the Safwan video aired. Attempting to prove that occupation was not the objective, Dari Alexander claimed that the United States was "truly trying to liberate." John Kasich was sure that liberation had already begun; he said, "Frankly, I think it's great people are being liberated" (Fox News Channel March 22).

One Fox guest, a military analyst, broke the monotony of the narrative and complicated it in a way that Fox journalists refused to do. He said, "We tended for some reason to think we were liberating the Iraqi people, and hopefully we are. We're invading a foreign country." He added that if Iraqis had invaded the United States, "some of us would fight" (Fox News Channel March 24), implying that it could be assumed that some Iraqis were resisting the invasion. The military analyst's brief interjection asserted that the situation was more complex than news commentators had portrayed.

Because CNN and Fox News Channel personnel positioned the troops as heroes battling a tyrannical regime to liberate an oppressed people, it was crucial to the war narrative to have evidence that the Iraqi reception of coalition troops was positive. CNN news staff circulated the Bush administration's belief that grateful Iraqis would cheer and welcome coalition troops. Anderson Cooper predicted that troops would be "greeted by adoring crowds," especially in Basra, where the reception would include "open arms . . . and praise" (CNN March 21 and 22). Wolf Blitzer and Aaron Brown also repeated the prediction that the reception in Basra would be positive (CNN March 21). Reporting on the view from the Pentagon, Barbara Starr conveyed the assumption of officials that U.S. forces would find friendly reception in southern Iraq, while Walter Rodgers reported that the commanding officer of the 7th cavalry expected more friendly Iraqis than hostile ones (CNN March 20). Similarly, on Fox News Channel, former assistant secretary of state Martin Indyk expressed his hope that Iraqis would cheer the U.S. tanks. Thomas McInerney was convinced that "jubilation" in Basra was imminent, a sentiment with which Simon Marks concurred repeatedly (Fox News Channel March 20). On-air news personnel established the expectation of gratitude with their own speculations and by repeating the assurances put forth by the Department of Defense. The next vital step in the unfolding of the marketable concept was the fulfillment of that expectation.

Embedded reporters provided the bulk of the information about coalition encounters with Iraqis, and those encounters were chiefly positive. When CNN reporter Martin Savidge was allowed to give an account of his journey into Iraq, he reported seeing Iraqis "waving at the convoy." When Carol Costello pressed him for more information, Savidge responded that the Iraqis "if not relieved, were at least not angry" to see the soldiers. In subsequent reports that day, Savidge relayed sights of "white flags," "thumbs up," and "waving" near Basra (CNN March 21). CNN journalist Ryan Chilcote also encountered Bedouins "waving," and Walter Rodgers described Iraqis "jumping up and down" (CNN

March 21 and 22). Later, Chilcote described a "very, very positive" experience with "lots of friendly contact" and an initial "warm" reception (CNN March 22 and 23).

Fox journalists also noted positive contact with groups of waving Iraqis; embedded reporter Greg Kelly acknowledged their presence twice and expanded on his experiences several times (Fox News Channel March 21). He described "very friendly" civilians who were "welcoming us" (Fox News Channel March 21 and 22). He went on to describe "Iraqi men and women" who were "friendly" and "receptive" with "genuine smiles" on their faces. He also described situations in which Iraqis gave them "the thumbs up." Later he elaborated, saying the Iraqis were "so warm," "seem[ed] happy," and did not hold "grudges" (Fox News Channel March 22). Significantly, Kelly added that he had not actually *spoken* to any Iraqis because no translators were available. The language barrier obviously hindered any meaningful exchanges between the reporters and the Iraqi citizens. Karen Andersen views this reluctance to engage with citizens as a deliberate strategy by U.S. reporters overseas; if the reporters do not understand the story, then there is no story to tell (2006, 71). Most U.S. citizens, including embedded reporters, have little or no knowledge of Arabic, a lack of knowledge that Edward Said describes as evidence of overwhelming resistance in the United States to understanding or learning about Islamic religion and culture (1997, xxxvi). So the speculative interpretations continued unabated, with CNN and Fox reporters working under the assumption that facial expressions and body language could easily be substituted for spoken language. In the studio, Juliet Huddy claimed that Fox News Channel was "hearing stories of civilians dancing" and "women . . . crying with joy," to which embedded reporter Kelly replied, "I wouldn't go that far" (Fox News Channel March 22). The repeated reports of waving Iraqis and celebrations at Safwan prompted Rita Cosby to say, "Fortunately, there's a lot . . . who are welcoming our U.S. troops" (ibid.). Colonel David Perkins, with whom Kelly was embedded, said that the locals were "very happy" and even "offered to sacrifice a goat and invite us all to dinner" (Fox News Channel March 23). Even through March 23, reports from the field were generally optimistic about the reception of the coalition in Iraq.

In spite of those reports, news of Iraqi resistance to the invasion eventually surfaced. Gary Tuchman, who was embedded with the U.S. Air Force, reported no resistance, while Aaron Brown announced there was "little resistance" (CNN March 21). Early on, Greg Kelly repeatedly issued reports of "no resistance" (Fox News Channel March 20 and 21). However, late on March 20 and March 21, when the military finally allowed many embedded reporters to transmit live, reports of resistance slowly began to appear on both networks.

At the Department of Defense, General Myers acknowledged on March 21 that there had been "sporadic resistance" around Basra but was quick to reiterate that it was limited (U.S. Department of Defense 2003d). That day, CNN Pentagon correspondent Chris Plante also reported "some resistance," and Christiane Amanpour reported resistance in the Iraqi port city of Umm Qasr (CNN March 21). At CENTCOM (United States Central Command) in Qatar, Mike Tobin reported "light" resistance, and in the studios at home Major Garrett, Gregg Jarrett, Jon Scott, Jeff O'Leary, and Steve Doocy repeated that report throughout the day. One report of "stern resistance" at Umm Qasr and another of "a lot of resistance" in Basra disrupted the consensus that Iraqi resistance was minimal. More reports that the resistance was only periodic toned down the impact of the reports of greater disruption, and Jon Scott said that what "small pockets of Iraqi resistance" existed would be under control within twenty-four hours (Fox News Channel March 21).

Later on March 21, William La Jeunesse reported that the U.S. troops came up against mortar fire and other artillery on the way to Basra. Tony Snow also announced that the coalition had been held up in southern Iraq because of resistance. Significantly, Fox News once again aired reports of celebrations at Safwan in between these two reports of resistance that was strong enough to engage U.S. troops. After Snow's report, two more stories about the celebrations at Safwan aired, followed by five reports of resistance in southern Iraq, Basra, and Nasiriyah. Two reports that resistance was nonexistent and two more re-airings of the reports of celebrations at Safwan followed those stories. March 21 ended with differing views on the resistance, from Chris Kline's report of "a lot of retaliatory Iraqi fire in Mosul and Kirkuk" to Oliver North's report of "significant" resistance to Bret Baier's report of "light" resistance based on information from the Pentagon (Fox News Channel March 21).

On March 22, reports of resistance were still inconsistent. Christiane Amanpour had heard news of "pockets of resistance" at Umm Qasr. Meanwhile, embedded reporter Jason Bellini experienced more resistance than his marine hosts "were anticipating," but Walter Rodgers did not categorize the resistance his particular group had encountered as "serious," although he did concede that the Iraqis were "not rolling over and playing dead." Ryan Chilcote, who had experienced a "very positive" encounter with civilians during the day on March 21, characterized the silence of that night as "eerie." Back in the studio, Anderson Cooper noted that fighting had replaced the warm reception that was anticipated at Basra (CNN March 22). Though March 22 began on Fox News Channel with Greta Van Susteren claiming there was "virtually no resistance so far," every subsequent report that day acknowledged "little" to "some" resistance that was "causing problems," but "not major problems" (Fox News Channel March 22).

When the Iraqi opposition continued into March 23 and March 24, Wolf Blitzer characterized the resistance as "stubborn" (CNN March 24). The self-assured belief that the Iraqis would indeed "roll over and play dead" was in evidence on Fox News Channel also, where Rick Folbaum described the resistance fighters at Umm Qasr as "pesky guys who seem intent on putting up some kind of fight, [though it was] unclear exactly why" (Fox News Channel March 23). Nevertheless, reports of resistance increased on March 23 and March 24, as did the intensity of the resistance.

Ruptures in the war narrative began to appear as the days wore on. CNN's Christiane Amanpour remarked that the anticipated warm reception and surrenders had not "happened to the extent that [the Bush administration] thought" (CNN March 24). Fox's Robert Maginnis also noted that he did not see the Iraqis "lining up on the streets cheering our arrival" (Fox News Channel March 23). Another military analyst at Fox observed that the Iraqis were not "throwing flowers at us," and Greta Van Susteren said, "It doesn't seem yet that we've gotten the enthusiasm from the Iraqi civilians" that was expected (ibid.). Trying to salvage the situation, David Hunt replied, "Because there's not a parade . . . in Basra . . . doesn't mean we haven't done the job right" (ibid.). By the end of March 24, the language with which resistance was described had changed considerably, with "little" and "some" resistance replaced by "stiff," "significant," "considerable," and "heavy" resistance. David Asman clung to the narrative, however, stating, "Some Iraqis have thanked our troops as liberators, while others have turned on them and fired" (Fox News Channel March 24). By positioning the Iraqis who resisted as traitors to the cause of the United States, Fox News Channel sought to massage the evidence to support the war narrative.

To bolster the faltering narrative, a considerable amount of rewriting ensued, and both CNN and Fox News Channel personnel began to report the fact that the Pentagon had repeatedly emphasized that the U.S. government and military had anticipated Iraqi resistance from the very beginning. Although before and during the first five days of the invasion Deputy Secretary of Defense Paul Wolfowitz, Secretary of Defense Rumsfeld, Assistant Secretary of Defense for Public Affairs Victoria Clarke, and General Richard B. Myers had consistently refused in press briefings and interviews to predict how quickly the war would conclude because of "uncertainties," CNN and Fox News Channel personnel had not heeded these cautionary words (U.S. Department of Defense 2003b). They had reiterated only part of the message—the conviction that a U.S. victory was inevitable.

The Safwan video simultaneously reaffirmed and called into question the liberation element of the war narrative. After being sidelined by the administration and the networks during "shock and awe," Iraqi civilians reemerged as major players in the story with the release of the Safwan video. They were valuable because they had no voice and CNN and Fox News Channel news personnel were free to speak for them, although Fox journalists employed this tactic more often than personnel at CNN. News of increased Iraqi resistance complicated and contradicted what the Safwan video appeared to show, and the networks altered their story to accommodate the changing interpretations of the mood of the Iraqis. The new narrative asserted that not all Iraqis were to be trusted, resistance was to be expected, and no one said the war was going to be easy. CNN and Fox reporters covered the resistance in the first five days only hesitantly, and they reiterated their commitment to a victorious outcome through their coverage of three events that threatened to rupture the narrative of liberation: the flag-raising at Umm Qasr, the discovery of a suspected chemical weapons plant at Al-Najaf, and the Iraqi capture of U.S. POWs.

### Umm Qasr

When the coalition attempted to take Umm Qasr—an Iraqi city with two ports, one older and one more recent—CNN and Fox news personnel continuously stressed that the city was a key portal for humanitarian aid. They also noted the importance of Umm Qasr for military strategy and for shipping in "heavy weaponry," but they mentioned this function less often than they discussed aid shipments (Fox News Channel March 21). Consequently, in the networks' reports, Umm Qasr's relevance for military strategy appeared to be tangential to the issue of humanitarian aid. Fighting at Umm Qasr began early in the invasion and continued unabated. In order to make this development work in the narrative, CNN and Fox journalists blamed the resistance for the fact that aid had not yet reached Iraqi civilians (CNN March 24; Fox News Channel March 22). Once again, the two networks placed the Iraqi civilians at center stage when their plight was necessary to boost the image of the coalition and reinforce the war narrative.

The coverage of the resistance at Umm Qasr was schizophrenic at best. CNN and Fox News personnel needed to control news of the resistance in order to preserve the integrity of the marketable concept. At approximately 2 PM Eastern Time on March 20, CNN revealed a Kuwaiti News Agency report that the coalition had taken control of Umm Qasr but noted that the British would not confirm the story. Fox coverage the same day reported simply that "U.S.-led troops" had taken Umm Qasr. The remainder of the coverage by the two networks, though essentially similar, retained the type of discursive difference that appeared during the release of the Safwan video. For example, CNN personnel handled news of the Umm Qasr victory gingerly at first; Lou Dobbs would not declare immediately that the coalition had taken Umm Qasr (CNN March 20). In contrast, Fox News Channel journalists instantly took to the idea. William La Jeunesse elaborated on the details the military had provided by saying that the troops "simply swept through" Umm Qasr, later adding that the city was "taken over very quickly by U.S. forces." Chuck Nash identified with the military as he announced the victory: "We've taken the town of Umm Qasr" (Fox News Channel March 20).

The next day, though, CNN and Fox News Channel reporting on the story took a number of confusing turns when studio personnel were too zealous in their embrace of the narrative. CNN's Anderson Cooper clarified that the coalition had taken only the new port of Umm Qasr and not the entire city, and Christiane Amanpour acknowledged "some resistance" at Umm Qasr, while at Fox, Brian Wilson stated that Umm Qasr was "still in Iraqi hands." When CNN's Carol Costello proclaimed that the battle at Umm Qasr was "pretty much over," her colleague Nic Robertson immediately contradicted her statement. Similarly, when Fox's Brian Kilmeade announced, "We could have a major . . . port city, in fact we do," Steve Doocy clarified that the port was "still occupied" by Iraqis. When CNN's Paula Zahn declared success at Umm Qasr, military analyst and retired general Don Shepperd clarified that the situation at Umm Qasr was "not all tied up" (CNN March 21; Fox News Channel March 21).

After this initial barrage of confusion, CNN and Fox adhered to one interpretation of the event. Paula Zahn reiterated the coalition's "very big success" at Umm Qasr and early on March 22 assured viewers that coalition forces had "wrapped up" security at Umm Qasr (CNN March 21 and 22). From 10:04 AM to 10:56 AM Eastern Time on March 21, Jon Scott repeated no less than five times that Umm Qasr belonged to the coalition. At the Department of Defense briefing on that day, Rumsfeld disputed the Iraqi information minister's claim that the Iraqi military had never lost control of Umm Qasr, stating, "In fact, coalition forces did capture it and do control the port of Umm Qasr, and also a growing portion of the country of Iraq." Later in the briefing, General Myers, too, affirmed that the 1st Marine Expeditionary Force and British troops had "secured" the city (U.S. Department of Defense 2003d).

Perhaps taking its cues from the Department of Defense, Fox News Channel maintained a similar position well into the early morning hours of March 22. Early that day, Brian Wilson challenged the Iraqi information minister's claim.

Wilson remarked, "He's making the case that Umm Qasr is not under British control. . . . We know that to be the case. . . . We've seen the pictures." Wilson's reliance on images as a source of truth came into question when Gregg Jarrett stated that Abu Dhabi TV had "suggested that Umm Qasr had not fallen" (Fox News Channel March 22). Meanwhile, with the exception of Amanpour's report of pockets of resistance north of Umm Qasr, CNN personnel did not waver from the news that the city was under coalition control. Although Amanpour reiterated on the afternoon of March 22 that Umm Qasr was "not fully fully fully pacified," three more reports that Umm Qasr had fallen ensued (CNN March 22). To make matters more confusing, CNN journalists relayed an Al Jazeera report that fighting in Umm Qasr continued. On Fox News Channel, one report that Umm Qasr had "been taken" followed another report that the coalition did not have control over the town (Fox News Channel March 22). In the Department of Defense briefing late on March 22, Victoria Clarke stated, "Coalition forces have the key port of Umm Qasr" (U.S. Department of Defense 2003e). However, Fox continued to air reports that Umm Qasr was "not totally in coalition control" (Fox News Channel March 22).

Finally, on March 23, CNN and Fox News Channel news commentators began to redefine the terms of their narrative. A shift occurred that day that amounted to another adaptation of the war narrative to unexpected events. The shift began when CNN's Amanpour ambiguously stated that Umm Qasr was "secure but not safe" (CNN March 23). Fox News Channel personnel and guest commentators began the process of redefinition; military analyst James Carey stated that the containment of a city by an invading force did not necessarily mean that the military had fully eradicated resistance. Shortly thereafter, Major Garrett explained that when the Pentagon said a town had "been taken," a new definition applied. According to the Pentagon, troops attempted to secure as much of a town as possible. If resistance remained, the troops tried to negotiate for surrenders. If that failed, troops would "engage" the combatants. Garrett said that "in a traditional sense, when cities are captured . . . that doesn't necessarily mean in the hours ahead there won't be some type of military engagement." By educating viewers about military tactics, strategies, terms, and definitions, Garrett attempted to recoup a narrative that was threatening to fall apart under the strain of two competing elements: conflicting information about the fighting at Umm Qasr and a general and simplistic sense of optimism on the part of network journalists. The networks' coverage of Iraqi resistance at Umm Qasr was key to maintaining elements of the marketable concept they were promoting—not just the superiority of the U.S. military but also the Iraqis' desire for liberation.

One specific incident at Umm Qasr further challenged the idea that the U.S. troops were liberators. In the midst of the back-and-forth that occurred on CNN and Fox News Channel early in the Umm Qasr struggle, U.S. Marines raised the U.S. flag over the new port. This act was perhaps one reason for the confusion over who exactly controlled Umm Qasr. CNN and Fox News Channel commentators reported the hoisting of the U.S. flag shortly after 4 AM Eastern Time on March 21. The raising of the flag invoked the image of occupying forces, an interpretation that CNN, Fox News Channel, and the Department of Defense repeatedly stressed was to be avoided. The U.S. and coalition leaders wanted the Iraqis (and the world) to regard them as liberators, not occupiers, and the presence of a U.S. flag on the battlefield suggested otherwise. Both networks needed to strengthen the liberation element of the war narrative while deemphasizing such impromptu and damaging images.

The flag-raising at Umm Qasr was not the first compromising act perpetrated by U.S. soldiers; a similar act had occurred the previous night. On March 20, Fox News Channel obtained footage from one of its embedded reporters of a U.S. soldier taking down an Iraqi flag that flew over a former UN post. Shepard Smith quickly contained the incident within the narrative and adapted the potentially disruptive image to the marketable concept:

> It goes without saying that one of the main goals of the coalition forces is to return to the people of Iraq their flag, their liberty, and their money . . . so this image of the flag coming down may be even more powerful at some point when we get to see the Iraqi flag go back up without the tyrannical leadership of Saddam Hussein. (Fox News Channel March 20)

Smith's statement perfectly adhered to the part of the marketable concept that stated that the United States was an altruistic liberating force.

Although CNN did not report on this incident, most likely because a Fox News embedded reporter figured largely in the coverage, CNN's studio journalists reported the Umm Qasr flag-raising and showed still photos of the flag without much editorializing. The extent of the commentary at CNN was limited to their description of the U.S. flag as the "Stars and Stripes," a mildly patriotic but factual term. Fox News, in contrast, offered a great deal of commentary. Rita Cosby broke the news at Fox News and commented on it with obvious pride and joy. In her initial report, she said that the marines had raised "the Stars and Stripes flag over" Umm Qasr, describing the event as "pretty incredible." She continued, "It is actually waving proudly there thanks to the U.S. Marines." A moment later, Cosby contrasted the flag-raising with the antiwar demonstrations occurring around the world, noting that "even though we saw an upside-down flag in the protests, we saw an upright flag" at Umm Qasr. A few minutes later, Cosby's commentary descended into giddy patriotism. Smiling, she remarked, "Who would have thought, in Iraq [giggles], many months ago that we'd ever see an American flag? Of course we're going to see a lot more of them in the coming days." She cited the occasion of the flag-raising as "obviously a happy time for the marines out there" as well as for their supporters in the United States. Fox's Adam Housely informed viewers that U.S. soldiers planned many similar gestures of patriotism as they proceeded through Iraq (Fox News Channel March 21).

Shortly thereafter, the positive outlook on Fox News Channel was challenged when E. D. Hill questioned whether the flag-raising was a "wise thing to do because we're there to liberate, not to occupy." She quickly settled the debate herself: "We had to do that," she said, because the Iraqi information minister had said that the United States "had not stepped one foot onto Iraqi soil." She concluded, "By raising the American flag . . . people . . . would see . . . something is changing here" and "that the information minister is lying." Her rationalization was similar to Smith's rationalization during the first flag-related incident, but she extended hers to include references to truth, proof that the U.S. troops were a liberating force, and regime change. Steve Doocy interpreted the flag-raising as "the kind of picture the administration has been waiting for," but Bill Kristol interjected that the soldiers would lower it eventually because "the point is to have a free Iraqi flag flying" (Fox News Channel March 21).

Roughly five hours later, CNN told viewers that the marines had lowered the flag. The two networks analyzed the lowering of the flag in different ways. Their reporting strategies salvaged the war narrative but demonstrated differing degrees of commitment to its liberation element. At 9:43 AM Eastern Time, CNN's Paula Zahn reported that the marines had lowered the flag; half an hour later, she explained that the Pentagon had ordered the action out of respect for the local Iraqi citizenry (CNN March 21). CNN journalists reported the lowering of the flag five times from March 21 through March 22. They reported the chain of events in a way that followed the Pentagon's narrative: U.S. Marines conquered resisting Iraqis at Umm Qasr, which prompted the exuberant soldiers to raise the U.S. flag, but the Pentagon ordered the flag lowered out of respect for Iraqi citizens. CNN's coverage of the flag-lowering merely repeated the rationale the Pentagon offered.

At 10:16 AM the Fox network made the first of six vague statements that indirectly touched on the issues surrounding the flag raising. Brigitte Quinn said, "It seems that we are being very, very careful not to cast the taking of these cities as the big conquest with the raising of the American or British flags, that, uh, it is an operation to liberate the Iraqi people." Retired army major general Robert Scales and Fox journalists Bret Baier, Shepard Smith, and Brit Hume offered variations on this theme, each wholeheartedly supporting the liberation aspect of the narrative. Robert Scales mentioned the feelings of the Iraqi citizens with his remark, "This is a war . . . not to conquer but to liberate . . . without alienating the population. . . . When you capture a center like we're doing right now . . . we have to project an image that this is a liberation, not a conquest" (Fox News Channel March 21).

Bret Baier framed his report in terms of the Pentagon's concerns. He reported,

> Senior defense officials say, "Listen, that is not the image that we want to show." . . . There are these soldiers that are doing great work . . . eighteen-year-old kids, some of them . . . but commanders are having those discussions with their troops because obviously pictures say things, and they don't want to send the message out that the U.S. and Great Britain are trying to conquer cities. They are liberating cities, and that is the policy, and that is what the U.S. and allied forces are doing. So, you are going to see some of these pictures. . . . They are making sure that the message is clear that this is a liberation of Iraq, not a conquering of Iraq. (Fox News Channel March 21)

Baier did not report the Pentagon's order to lower the flag. His narrative emphasized the impulsive acts of youth and the incorrect messages that the images those acts created might convey.

As if to exercise more damage control, Quinn commented that Fox News Channel had "been very emphatic in casting this as the liberation of Iraq," thus betraying the heavily constructed nature of the coverage. Shepard Smith said, "There is the hope, of course, that in the weeks ahead . . . they'll be able to return the Iraqi flag and return this country to the Iraqi people" (Fox News Channel March 21). As with his remark about the first flag-related incident, Smith circumvented the specter of occupation and focused instead on what the absence and possible reappearance of the Iraqi flag meant to the Iraqi people. Brit Hume announced that the Pentagon was "not encouraging the practice" of flag-raising and said that the British forces had been "forbidden" from raising their flag "on the grounds that the hoisting of the Union Jack might smack of occupation and conquest, not liberation and freedom" (ibid.). Hume's remark was the last of the day that touched on the issues surrounding the flag raising without actually mentioning the flag's lowering.

The day after the flag was lowered, Gregg Jarrett twice mentioned the flag-raising at Umm Qasr without mentioning the fact that the Pentagon had ordered it to be lowered. Later in the day, Mike Tobin, repeating an exchange with

General Tommy Franks at CENTCOM about the incident, said that the marines had lowered the flag in recognition of the U.S. goal of liberating the Iraqis (Fox News Channel March 22). That was the first mention on Fox News Channel that the U.S. flag had been taken down. Though Fox personnel diverged from the Department of Defense's straightforward explanation of the situation, their refusal to report the marines' faux pas demonstrated a greater degree of commitment to the overall message of the U.S. government and military.

The networks' treatment of the resistance and the flag-raising incident at Umm Qasr supported their simplified war narrative; when evidence threatened the veracity or the plausibility of the narrative, they massaged the evidence to make it fit or deleted it altogether. For the journalists and embedded reporters at CNN and Fox, the war narrative was unfolding as planned, in spite of the continuous fighting and questionable images. They interpreted the two flag incidents in a way that solidified the liberation aspect of the war narrative, illustrating their need to keep the narrative intact and safe from unsettling images. Moreover, Fox News personnel used the two flag-raising incidents to speak for Iraqi civilians. In their narratives, they positioned the Iraqis in a variety of ways to serve a multiplicity of purposes within an ideologically unified framework. That framework, unfettered by the complications of the invasion and bolstered by the statements of the Department of Defense, was sustained by the networks' continuing simplification of events.

### Al-Najaf Chemical Weapons Facility

In the moments preceding and following the toppling of the Hussein statue in Al-Firdos Square on April 9, CNN and Fox News Channel reinforced one theme that guided the war narrative: the Hussein regime was a threat to its own people and to the United States. On that day in April, the weapons that would prove that Iraq was a threat to the United States still had not been found. Two weeks earlier, the urgent need to find those weapons and the possibility that they had been discovered led CNN and Fox News to intensify the disarmament portion of the narrative through the creation of a credible threat.

Support for the Bush administration's original war narrative required agreement with its basic premise that the Hussein regime's WMDs posed a threat to the United States. The Bush administration accused the regime of stockpiling WMDs and having ties to Al-Qaeda and claimed that that combination yielded the probability of another terrorist attack on U.S. soil. The administration's primary argument, then, was a major component of the marketable concept: the vengeance the United States would exact was a safeguard against future, possibly larger-scale, attacks. The fervor with which CNN and

Fox News personnel perpetuated this premise, despite evidence to the contrary, was a testament to their devotion to the guiding ideological framework and the simplistic narrative it produced. The networks' ability to create the Iraqi threat and construct it in accordance with the administration's narrative depended on a representation of Iraq's arsenal as a global hazard. The discovery of a potential chemical weapons facility in Al-Najaf on March 24 provided the opportunity for such a depiction. CNN and Fox News spent a great deal of time on the story, relentlessly repeating their suspicions to defuse the contradictory components of the discovery.

Long before the find, the Department of Defense presented Iraq's chemical and biological weapons as a given. In a briefing on March 11, Rumsfeld stated, "We know he continues to hide biological and chemical weapons, moving them to different locations as often as every twelve to twenty-four hours, and placing them in residential neighborhoods" (U.S. Department of Defense 2003a). On March 12, Wolfowitz reinforced Rumsfeld's statement: "I still think the thing that worries me most is the use of chemical and biological weapons. We're quite sure he has them" (U.S. Department of Defense 2003b). General Myers assured the press on March 21 that the U.S. "will disarm the Iraqi regime and ensure their weapons of mass destruction will not fall into the hands of terrorists" (U.S. Department of Defense 2003d). In his statement, Myers not only articulated the threat but also justified the preemptive strike by mentioning terrorists, an unmistakable reference to the September 11 attacks. The claims of the Department of Defense went virtually unchallenged by CNN and Fox News Channel journalists; they simply adopted the department's case.

Ten themes dominated discussions of WMDs on the two networks: the certainty that WMDs existed; the number of possible WMD sites in Iraq (CNN said 1,400, Fox News Channel said 650); the ongoing search for WMDs; the possibility that the search would take a long time; the implications if WMDs were not found; the possibility that WMDs were hidden; the chance that Hussein would order the use of WMDs if pressured; the hope that WMDs would not be used; the possibility that Iraqi soldiers were armed with WMDs; the coalition bombing of WMD-related targets; and the fact that the discovery of WMDs would vindicate U.S. actions. Only one theme involved a scenario in which WMDs did not materialize, but the remaining nine assumed that the existence of an arsenal of WMDs was a foregone conclusion.

CNN and Fox News personnel discussed Iraq's arsenal at great length, offering estimates of its weapons capacity that were even greater than that of the Department of Defense. For instance, although commentators at CNN and Fox emphasized that Iraqi missiles were inaccurate, the alleged presence of

Scud missiles in Iraq in the first few days of the invasion prompted Fox News to construct the Hussein regime as noncompliant with UN Security Council resolutions. This created the notion that there was an implicit link between the presence of Scuds and a stockpile of WMDs. On March 22, Adam Housely said that Iraq had fired "at least seven" missiles at Kuwait (Fox News Channel March 22). Fox News repeatedly stressed that the Scud missiles Iraq allegedly possessed violated UN Security Council resolutions. William Cowan alluded to a probable WMD stash when he said that Hussein had "done a masterful job of hiding" the Scuds. After that, the language shifted subtly from claims that the Scuds (of which there was no evidence at that point) were illegal to the assertion that "Saddam said he did not have" these weapons (Fox News Channel March 20). Jon Scott remarked, "Once again, proof that Saddam Hussein has been lying through his teeth," implying that if Hussein was lying about the Scuds, he was also lying about Iraq's development and possession of WMDs (Fox News Channel March 21).

Fox News Channel's repeated assertions that Iraq had or had fired Scud missiles were not based in fact. From March 20 through March 23, Fox claimed that Hussein had lied when he said that he did not have Scuds, even though on March 22, the Department of Defense released a statement that contradicted Fox's claims that Hussein had lied. Bret Baier reported that the Pentagon had "admitted" that no Scud missiles or Scud launchers had been found and that Iraq had not fired any Scuds, though teams were "searching frantically" for evidence to the contrary (Fox News Channel March 22). Indeed, General Mc-Chrystal stated in a Department of Defense briefing on March 22 that the Iraqis had not launched any Scuds and that the coalition had not found evidence of Scuds on the ground (U.S. Department of Defense 2003b). Nevertheless, Adam Housely reported again on March 23, as he had the day before, that some of the missiles Iraq had fired at Kuwait were Scuds. On March 24, Housely reported that the U.S. military had said that Iraq had not fired Scuds but that Kuwait maintained that two missiles were in fact Scuds. That afternoon, Major Garrett reported from the Pentagon that Iraq had not fired Scuds at Kuwait. Fox News Channel journalists finally retracted their accusations late on March 24, but that did not retroactively eliminate their repeated claims that Saddam Hussein had been deceitful. The Scud discourse at Fox News organically reinforced the WMD discourse. Only one Fox employee dissented from this narrative; Greta Van Susteren pointed out that the presence of Scuds did not indicate the presence of WMDs (Fox News Channel March 22).

Meanwhile, CNN and Fox News Channel kept the possibility of chemical and biological weapons in the forefront of viewers' minds, albeit with negative evidence. As journalists followed every missile fired into Kuwait from Iraq, they notified viewers that the warheads contained no traces of chemical or biological weapons. At the same time, reports from embedded journalists repeatedly addressed the need for soldiers and journalists to don heavy protective gear in case of a chemical weapons attack. On March 20, Chris Plante reported that a plane that crossed into Kuwait might have been carrying chemical weapons; the U.S. military had told him that Iraqis commonly used civilian planes as "chemical weapons crop dusters" (CNN March 20). CNN on-air personalities repeated the story seven times that day.

The story that CNN and Fox News constantly returned to was the rumor that the Iraqis had drawn a "chemical weapons red line" around Baghdad. Supposedly, when U.S. soldiers crossed this line, the Iraqi Republican Guard would unleash their chemical weapons. The best examples of the tone of the discourse are Gregg Jarrett's statement at Fox that "*so far . . .* no chemical or biological attacks," and a crawl on CNN that read "Pentagon: No weapons of mass destruction found *yet*" (Fox News Channel March 21; CNN March 22; italics added). Journalists at both networks cultivated a sense that these attacks were imminent or, as Greta Van Susteren put it, "likely" (Fox News Channel March 24). Fox journalist Eric Shawn asked military analyst Wade Ishimoto if the reason Iraq had not used chemical weapons was because coalition special operations forces had entered the country very early in the invasion. Ishimoto answered in the affirmative, thus allowing for the possibility that the absence of chemical weapons attacks did not signify the absence of chemical weapons altogether (Fox News Channel March 23).

The discovery of a suspected chemical weapons facility reenergized the networks' construction of Iraq as an unusually deadly threat. On March 23, the *Jerusalem Post* reported that the United States had taken a suspected chemical weapons facility in Al-Najaf in southern Iraq. Immediately after the discovery was made, Wesley Clark said the find would be "smoking gun evidence" if weapons were located. CNN's senior Pentagon correspondent Jamie McIntyre reported that the "alleged chemical factory . . . may not look quite as exciting by tomorrow" but said that it "may . . . have produced chemical weapons at some time" (CNN March 23). Clark and McIntyre did not dismiss the possibility that the facility was active; to have done so would have seriously compromised the discourse about WMDs. CNN's Carol Costello was not as reticent as others were on her network; she asserted that the facility was camouflaged and booby-trapped (CNN March 24).

The analysis of Fox journalists contrasted sharply with that of most CNN personnel. They claimed at the outset that the United States had discovered a "huge chemical weapons facility" and said that it was a "big find," "a homerun," "pay dirt," and "egg on the face of UNMOVIC [United Nations Monitoring, Verification, and Inspection Commission]" (Fox News Channel March 23 and 24).[3] Steve Doocy and Adam Housely concurred with this interpretation, commenting respectively that the facility "smell[ed] like a weapons of mass destruction plant" and had "all the earmarks of a chemical facility" (Fox News Channel March 24). Although CNN had used the word "possible" to describe the plant since the discovery was made, Fox News journalists dismissed such caution. David Asman said, "It doesn't sound like a baby formula company there," and Tim Trevan later commented, "It's not like it's a baby aspirin factory" (Fox News Channel March 23 and 24).Fox News journalists were certain that the find was legitimate vindication for the invasion.

They relinquished this interpretation with great reluctance. However, on the night of March 23, a note of qualification crept into Fox's reporting of the discovery. Bret Baier stressed that Pentagon officials were calling the Al-Najaf facility a "suspected" chemical weapons plant (Fox News Channel March 23). Just as CNN's Jamie McIntyre had predicted, on March 24, the U.S. military reported that it had found no chemical weapons at the facility. CNN's Barbara Starr reported that in spite of the uncertainty about the Al-Najaf site, military officials expected to find "other sites, other facilities that will have some relationship to Iraq's WMD program" (CNN March 24). CNN preferred to let the "discovery" of Al-Najaf site fade away and focus on probable future sites.

Fox News Channel took a decidedly different tack and resuscitated the story. Although Major Garrett conceded that the discovery was less of a story "than originally thought," he noted that the facility "did look to be capable" of weapons manufacturing and said that Hussein had ordered it evacuated recently. Soon after, Sean Hannity and Shepard Smith had an exchange that again underscored the degree to which Fox News Channel clung to a narrative that seemed to be falling apart. Hannity began by saying, "So, weapons from that chemical facility that they found in the south," but Smith interrupted and contradicted him, saying "Not sure." Hannity replied, "Well, reports are pretty clear." The narrative stayed alive at Fox by the sheer force of that type of dis-

---

3. Since December 1999, UNMOVIC has been responsible for ensuring that Iraq is in compliance with UN Security Council resolutions in an effort to prohibit its development of WMDs; see basic information about this UN operation at www.unmovic.org.

course. Immediately after the exchange, Smith infused more doubt into what was a clear-cut situation with his statement, "[The Iraqis] were disguising this thing for some reason." Greta Van Susteren remarked, "Chemical weapons are now . . . rearing their ugly head." Her statement implied that the Al-Najaf facility was teeming with WMDs despite the clear evidence to the contrary. Smith continued to perpetuate the idea that the military had indeed found WMDs by identifying Al-Najaf as the place with the chemical weapons plant; Bret Baier had to correct his mistake (Fox News Channel March 24).

The Al-Najaf facility revitalized the WMD discourse for a day, which can seem like an eternity on a repetitive 24-hour news channel. Up to the point when the facility was discovered, discussions about WMDs were full of conjecture even though neither CNN nor Fox News journalists wavered from the conviction that Iraq had and would use chemical and biological weapons. CNN personnel cited the Al-Najaf facility as potential proof of illegal weapons, while journalists at Fox asserted that the proof was definite. Once the Department of Defense killed the story, CNN personnel used the story to stress that the search for more sites was just beginning. Fox News Channel commentators chose to retain their skepticism about reports that the facility did not have WMDs. Despite their different approaches to narrativizing the discovery of the facility, both networks reported on it regularly because it was a compelling way to establish and sustain the idea of a plausible threat.

### U.S. POWs

On March 23, Iraqi vice-president Taha Yassin Ramadan announced that U.S. prisoners of war would soon appear on television. If his claim proved to be true, Ramadan's announcement would fly in the face of one of the primary constructs of the marketable concept—that the U.S. military was vastly superior to the Iraqi military. Although no one expected that the U.S. invasion would be free of casualties, CNN and Fox News Channel journalists reacted to Ramadan's announcement as though a small Iraqi victory was not even plausible. At Fox, Rick Folbaum denied the existence of any reports of captured coalition soldiers, Catherine Herridge cautioned viewers to take the information "with a grain of salt," and James Carey said he refused to believe the story (Fox News Channel March 23). Their reaction followed reports emanating from CENTCOM that no soldiers were missing. However, subsequent reports from Rumsfeld affirming the capture of U.S. soldiers shifted the discourse. The video of U.S. POWs shot by Iraqi TV and aired first by Al Jazeera gave CNN and Fox News journalists the material to recoup the narrative of heroism, superiority, and the Iraqi threat. They did this by downplaying a questionable

U.S. military strategy, placing blame on Iraqi citizens, accusing the Iraqis of mistreating the POWs, and emphasizing the coalition's magnanimous treatment of Iraqi POWs.

On March 23, the 507th Maintenance Company took a wrong turn in Nasiriyah. Iraqis who were assumed to be pacified by coalition troops that had already left the area ambushed the soldiers. As CNN and Fox News Channel journalists conveyed in their coverage of the Umm Qasr resistance, the Pentagon's definition of taking an area did not involve complete pacification. For that reason, resistance was to be expected. David Hunt defended the strategy, saying that resistance was no surprise and that bypassing was a successful method used in World War II (Fox News Channel March 24). However, this strategy left the maintenance crews that supported the long-departed troops vulnerable. At the Department of Defense, Victoria Clarke characterized the tactics of the Iraqis who ambushed coalition forces as "deadly deceptions" and "terrorism" (U.S. Department of Defense 2003f). Similarly, Hunt called the Iraqis who captured the POWs "bad guys" playing "tricks." In this instance, Hunt used the idea of Iraqis who were "fighting dirty" to supplant the notion that the U.S. military strategy of leaving troops behind might be questionable (Fox News Channel March 24).

Another strategy for handling this evidence that did not fit the marketable concept of a superior U.S. military force was to focus on Iraqi civilians. Suddenly they appeared to be a looming threat. In an interview with Air Force Secretary F. Whitten Peters, Fox's Kiran Chitry asked how a "secured" location like Nasiriyah could have been taken over by the very people who had been pacified. Peters replied that because the civilian population was not locked up, it was difficult to stop a "terror campaign" launched by motivated people. Chitry responded by asking if the coalition was taking "risks" such as not locking up civilians because the soldiers were "trying so hard to reduce civilian casualties" (Fox News Channel March 23). Victoria Clarke at the Department of Defense also drew civilians into the discussion, claiming that the Iraqi ambushes compromised the ability of the coalition "to accept surrendering forces or protect civilians" (U.S. Department of Defense 2003f). However, she did not say that protecting civilians hampered the ability of soldiers to do their jobs, as Chitry did.

Fox News journalists described Iraqi civilians as the ungrateful recipients of humane treatment. Sean Hannity repeated Chitry's question in an interview with Alexander Haig, "Are we putting too much emphasis on avoiding collateral damage? Are we putting our troops in harm's way to a greater extent? Because we don't want that" (Fox News Channel March 24). By asking the question and voicing the idea, Hannity transferred blame from a coalition strategy put into place to compensate for a dangerously low number of troops to the very Iraqi civilians that the United States was purportedly doing its utmost to protect in a "humane war." Fox News commentators preferred to question the strategic value of "taking extraordinary measures to prevent innocent casualties" rather than question the value of the military's bypassing strategy (U.S. Department of Defense 2003a).

The price of protecting civilians, as Fox News Channel's explanation posited, was the Iraqi capture of U.S. soldiers. As soon as it became clear that there were U.S. POWs, both CNN and Fox News devoted a great deal of time to discussing the Geneva Conventions. Personnel at both networks claimed that the fact that the Iraqi forces broadcast images of the POWs violated the conventions, though they defended the countless images of Iraqi POWs their networks broadcast, which were made possible by the U.S. military's embedding system (CNN March 23; Fox News Channel March 23).

Personnel at both networks also focused on the provisions of the Geneva Conventions concerning torture. Both CNN and Fox News Channel invited POWs from previous wars into their studios to discuss their experiences. Concern about the well-being of POWs was not without basis or unusual. The possibility that captured U.S. soldiers could be tortured or raped was legitimate and newsworthy, but the capture of these soldiers had the potential to disrupt the scripted narrative of a preordained and quick victory, and both networks moved into damage-control mode. The marketable concept presupposed that coalition soldiers would not become POWS. However, once the Iraqis captured some soldiers, network personnel did not need to work so hard to create a credible threat in their discourse; the threat to U.S. soldiers had become real. Significantly, by March 23, when Iraqis had killed and/or captured twelve soldiers, CNN and Fox News Channel had already reported the capture of thousands of Iraqi POWs, despite the fact that the networks did not have access to reliable statistics at that point. When juxtaposed with the abuse and murder of U.S. POWs, the U.S. military's treatment of Iraqi POWs made the narrative and marketable concept eminently believable.

In addition to showing images of Iraqi POWs, CNN and Fox News journalists discussed international law regarding their treatment and specific instances when coalition soldiers gave them fair, humane, and sometimes extraordinary treatment. Those anecdotes reinforced the notion that the invaders were selfless humanitarians, an important aspect of the marketable concept. For example, relaying what she had seen of the Iraqi POWs in coalition custody, Rita Cosby stated, "We . . . are treating their soldiers better than [the Iraqis] are." Reports that some Iraqi soldiers lacked basic necessities such as shoes and other reports of U.S. soldiers that "felt sorry for" them fed that type of claim (Fox News Channel March 22).

More dramatic were stories from both networks that played up the magnanimous heroism of the coalition troops. Late on March 23, Dr. Sanjay Gupta, who was embedded with a MASH unit, reported watching "the first . . . abdominal operation for a gunshot wound" of the war in a tent in the desert. A little over an hour later, Aaron Brown revealed that the victim was actually an Iraqi soldier. After a pre-edited story on the aftermath of a battle at Nasiriyah, Brown noted that on that day Dr. Gupta had "witnessed a surgery . . . on an Iraqi POW performed by American doctors" (CNN March 23). Fox News Channel had its own story of medical heroism. Greg Kelly told of an incident in which one member of his Fox News Channel crew, a former paramedic, helped soldiers treat Iraqi POWs injured in the previous night's firefight (Fox News Channel March 23). Such stories enhanced the overall narrative. Not only did the coalition troops make a genuine effort to shield Iraqi civilians from harm, but they gave wounded Iraqi soldiers top-notch medical treatment that was unavailable through resources in Iraq. In the narrative the networks offered, the selflessness of the troops and embedded journalists was but one more illustration of the selflessness of the entire operation.

## Conclusion

Though the war narrative seemed coherent and strong at the outset of the conflict, it had numerous weaknesses. Contradictions persisted throughout the coverage, as should have been expected in an undertaking of that size. The narrative could not continue and conclude as easily as CNN and Fox News Channel on-air personalities made it seem. In constructing the 2003 war as a high-concept narrative, CNN and Fox News Channel encountered holes in the plot and contradictions in the details. When this happened, they chose to massage their analysis of evidence to make it fit their narrative.

The appearance of U.S. POWs and soldiers who had been killed in action on television had the ability to dismantle a fragile narrative. The narrative had already shown signs of wear with controversial imagery of a U.S. flag planted in Iraqi soil, evidence of mounting Iraqi resistance, and the conspicuous absence of WMDs. Douglas Kellner calls the accumulation of events like these a "spectacle of chaos" (2005, 63). Before the situation reached that point, however, personalities on CNN and above all Fox News Channel tried a number of discursive strategies to preserve the integrity of the narrative. They redefined terms, chided overeager soldiers, manipulated inconvenient details, and elevated heroes to another level.

It bears repeating that high-concept films have never earned respect for their narrative cohesion or sophistication. After examining these six events, I can attribute the same flaws to the war narrative. Simplified narratives are inherently deficient, and as I reconstructed the narrative that CNN and Fox News Channel followed, I found that the deficiencies in the narrative did the greatest disservice to Iraqi civilians. News commentators imposed identities on Iraqi civilians that were not accurate, and these identities kept changing as the narrative required. They had no stable role in the war narrative. Fox News Channel in particular molded the Iraqi civilians to suit whichever twist arose that day. The alleged thoughts, sentiments, and actions of Iraqi civilians were invoked and manipulated to fit almost any scenario that bolstered the image of the coalition. The idea of liberating the Iraqis was important to the narrative, but this idea seemed to work best narratively when the Iraqis themselves did not figure into the storyline. As a result, U.S. network journalists made Iraqis absent and present as needs dictated and spoke for them when it served the narrative both networks followed. They did all of this in the service of the extremely simple and superficially uncomplicated narrative that "editors would accept and that the North American public would recognize" (Andersen 2006, 71).

One of the ways network journalists kept the narrative simple was by focusing on the aesthetics of the images they were receiving from Baghdad. They used the images of the "shock and awe" bombing campaign as an opening for a disproportionate amount of discussion about safe precise weaponry and the visual appeal of the explosions. The on-air journalists dedicated little time to other types of commentary or analysis. For them, the bombings were visually spectacular; indeed, "shock and awe" was pure ideological spectacle that eclipsed almost any notion of wounded or dead Iraqi civilians. CNN and Fox News Channel journalists participated in the spectacle with a focused and almost monolithic narrative discourse. In salvaging the war narrative from realistic complications and multiple viewpoints, they embraced the message of the Bush administration and the Department of Defense.

Beyond the plain fact that virtually no on-air personalities critically interrogated the government's policy, the narrative the networks constructed, adapted, and maintained couched the events in simplistic and decontextualized terms. The narrative discourse, particularly on Fox, was reductive and therefore politically regressive, relying on official views to guide—but not necessarily limit—the narrative. CNN remained true to the information that came from the Department of Defense and the White House. Fox on-air personalities may have taken their cues from Department of Defense briefings, but their discourse often led them beyond what even the Department of Defense was willing to express aloud and in public. It is important to remember that Fox News Channel's departure from the official line was in no way critical; instead, the on-air personalities simply clung to the marketable concept with more strength than CNN did. The marketable concept remained intact on both networks, and differing degrees of discursive enthusiasm did not alter the networks' ideological similarities—similarities rooted in the marketability of high-tech revenge, defense, and liberation.

The discursive differences did serve a purpose with regard to high concept, however. High-concept films rely on differentiation to pursue diverse segments of the audience. The tools used to differentiate high-concept films exist in television, and each television network has the advantage of a comprehensive brand name built over time. The fully branded channel is a holistic identity that can persevere with a loyal audience. Because CNN and Fox compete within the same audience pool—cable and satellite subscribers—their unique brand names are essential to their success in the cable news marketplace. Although the information each network imparted in the first five days of the invasion was similar, the ways that they narrativized that information spoke to the different strategies they employed to attract certain market segments. That type of differentiation in high-concept television is not specific to war coverage, and it is more apparent in long-term coverage.

# 3

## *Intertextuality, Genres, and Stars*

On March 23, 2003, Iraqi resistance fighters engaged with U.S. soldiers at a location that embedded reporter Walter Rodgers was not allowed to identify. Some of the Iraqis were riding in a pink pickup truck with an NYPD sticker on its window shield. The pastiche and marketable concept that the truck represented so impressed Aaron Brown that he remarked, "In the context of why this war is being fought, you come across a pickup truck with an NYPD sticker on it. . . . I mean, you can't write movies like this!" (CNN March 23). In Brown's estimation, the real was so overwhelmingly fantastic that not even Hollywood could have concocted it. Brown's allusion to the movies was not a one-time occurrence. On another occasion, the sight of camels was enough to compel embedded reporter Greg Kelly to exclaim, "It's like a movie!" (Fox News Channel March 21). These are just two examples, but they epitomize the way that many journalists referred to media artifacts and even historical events to contextualize or explain the events of the war. Ironically, these easily comprehensible references did the exact opposite; they used common understandings and memories as poor substitutes for realism and context.

Ease of communication is a vital component of marketability. In high-concept films, any product can be sold in the marketplace as long as its basic concept is easy to convey and easy to understand. After all, the goal of high concept is the sale of the film and every product that marketers can link to it. This basic tenet is true for high-concept television as well. In television, the network uses its programming to attract an audience that it can sell to advertisers. The network must entice the audience to remain with a particular program over the course of a season and even longer. In the case of wholly branded channels like the Food Network or CNN, the networks asks the audience to watch several related cooking shows or several hours of news programming. Even though the product for sale on television is primarily the audience, the programs generate other marketing opportunities: DVDs, clothing, companion books, and in some cases (as with *The X Files, The Simpsons,* or various children's programs) action figures, dolls, and video games. CNN has not diversified to that degree, but Fox News Channel is well on its way. Mugs, umbrellas, T-shirts, and other goods are available for sale on the Fox News Web site. Both networks exploit the commercial potential of their most marketable programs and/or coverage.

The marketability of television news, like that of films, relies most noticeably on the story and the visuals. Within the story are various components—intertextuality, genre, stars, and character types—that contribute to the programming's marketability by ensuring the simplicity of the message and reinforcing the formula. Formulas guarantee the marketability of the product by guarding against narrative anomalies. They impose strict boundaries and confine the life of a story. Such containment of information holds steady in all types of media, including news programming.

Even though live events covered in news programming could end in any number of ways, the pull of formula is unwavering. Susan Moeller argues that the liveness of news events is exciting precisely because of the events' apparent immunity to predictability. However, "once the parameters of a news story have been established," she writes, "the coverage lapses into formula" (1999, 13). As a result, the oft-repeated narrative of heroes who overcome obstacles to triumph over evil colors news coverage of crises, robbing it of context and uniqueness by superimposing reductive templates. The hero plot is one of "overt action," according to Sharon Lynn Sperry, and its formula relies on the myth of the "single individual of superior worth and superior skill, who will meet the problem and conquer the evil" (1981, 300). In slightly altered form, this myth informs the marketable concept of the 2003 invasion—a simple idea about U.S. technological superiority, altruism, and vengeance that evades the problematic issues of war. For John Fiske, the popularity of news programming relates directly to this propensity for using "generic characteristics" that confine "reality" (1987, 283). When formulas are present in television news, they are encumbered by events that are taking place in real time around the world. The liveness that television is suited to is a particular threat to the integrity of formulas. Because the simplicity of a formula cannot restrain the magnitude of a crisis in international relations, it attempts to destroy the complexity of such events, according to Moeller and Fiske.

Consequently, complexity yields to easy and quickly communicated stories that mimic the pared-down stories of high-concept films. As a means of fulfilling viewer expectations and limiting deviations from what has sold in the past, formula works in tandem with intertextuality to impart information swiftly and predictably. This chapter examines how formula and intertextuality converged in CNN and Fox News Channel's initial coverage of the 2003 invasion of Iraq. Intertextuality functioned in the war coverage to communicate reductive representations of events. The agents of intertextuality—genres and stars—performed likewise. CNN and Fox New Channel personnel relied upon them to decontextualize and sell the invasion of Iraq.

## Intertextuality in the 2003 War Coverage

Intertextuality is a common yet complicated process in which one text refers to other texts (White 1992, 162). For example, when the character of Maddie from the soap opera *As the World Turns* has a morbid daydream set to the theme song of the HBO drama *Six Feet Under,* viewers witness an instance of intertextual referencing. Viewers' knowledge of *Six Feet Under*'s macabre subject matter (the program is set in a funeral home) gives them a greater appreciation of Maddie's anxiety-producing daydream. Intertextuality can be used for parody, irony, or to reward knowledge of pop culture in meaningful or pleasurable ways. Intertextuality is one indication of the text's "hyperawareness . . . of its cultural status, function, and history, as well as of the conditions of its circulation and reception" (Collins 1992, 335). By referring to other texts, the hyperaware text "hijack[s]" its history and endows it with a new cultural significance (333). Consider the cultural and narrative benefits to the viewer of catching and correctly identifying all of the intertextual references on such television programs as *The Sopranos* and *The Gilmore Girls.*

In high-concept films, intertextuality acts as shorthand that plays to and relies upon the audience's vast knowledge of media artifacts to impart information quickly. High-concept films use intertextuality for economic reasons (Wyatt 1994, 58). If the audience can piece together narrative information with the aid of one quick reference, then filmmakers do not have to spend screen time explaining or portraying the information. Shorthand does more than serve an economic purpose in high concept. It also activates the audience's commonly held assumptions that play into the ideological aspects of high concept.

Intertextuality pervaded the war coverage of CNN and Fox News Channel. Network journalists not only referred to other media texts and figures but also to historical events as a short-hand way of discussing current events. In doing so, they sought to tap into the audience's knowledge of certain events, which meant that they believed that their understanding of history was the same as that of their audience. It is important to note that when network journalists referred to historical events, the referents were either mediated representations of the events (films, etc.) or discursive constructions of the events (how the events have survived in the mainstream U.S. imagination). They treated history as a stable entity rather than dealing with historical events as a complex set of social, political, and economic relations. In coverage of the first five days of the 2003 invasion, CNN and Fox News Channel used three major categories of intertextual references: references to past international conflicts, references to film genres, and references to movie stars.

### References to U.S. Military History

The networks referred to military history to stress the technological supremacy and moral authority of the U.S. military. References to World War II, the Vietnam War, and the 1991 Persian Gulf War propped up the marketable concept of the 2003 invasion—the notion that the invasion was a justified act of vengeance by a technologically superior U.S. military force that was in Iraq to liberate its citizens from the Saddam Hussein regime. Daniel Hallin (1994) has pointed out that the invocation of World War II typically emphasizes a proud wartime tradition and a clear distinction between good and evil. In the coverage of the 2003 war, invoking World War II carried inconsistent meanings. CNN and mostly Fox News Channel journalists referred to World War II for three primary reasons: to note similarities of World War II with the 2003 invasion, to note differences between World War II and the 2003 war, and to speculate about the actions of Saddam Hussein based on the actions of Adolf Hitler. All three drew from commonly understood interpretations of history to strengthen the marketable concept of the 2003 war and justify the invasion.

The references to similarities between the 2003 invasion and World War II were relatively few, but they tended to be positive and reassuring in order to emphasize the competence and humanity of U.S. soldiers. On Fox News Channel, retired general Thomas McInerney compared the push of the 3rd Infantry Division into Iraq to Patton's race to Paris (Fox News Channel March 21). Footage of U.S. soldiers throwing candy at Iraqis as they passed through a town prompted embedded reporter Greg Kelly to compare the act with "something . . . they used to do back in World War II" (Fox News Channel March 22). And retired colonel David Hunt justified the U.S. strategy of by-passing resistance by stating that it was the same strategy used in World War II (Fox News Channel March 24). These instances established World War II as a standard of warfare by which the on-air personalities measured the 2003 invasion.

The differences between World War II and the 2003 invasion that CNN and Fox News Channel journalists noted starkly contrasted with the similarities they noted. In these instances, network personalities positioned World War II as an outdated model of warfare and their references focused on precision bombing, speed, and strategy. These topics fortified the issue of technological superiority that was an important element of the marketable concept. Particularly on Fox News Channel, journalists made it their mission to differentiate between what David Hunt called the "humane bombing" of 2003 and the imprecise bombings of World War II. Some members of the Department of Defense embraced this distinction. In the March 21 Department of Defense briefing, Rumsfeld spoke at great length about how "no comparison" existed between the 2003 bombings and those of World War II (U.S. Department of Defense 2003d). Similarly, Fox News Channel's Bob Sellers attempted to minimize the ferocity of the images of "shock and awe" by saying, "It's not Dresden. It's not Hiroshima" (Fox News Channel March 22). The filmed images of both of those cities, especially Hiroshima, depict total annihilation—a state of utter disaster that Fox News Channel earnestly tried to separate from the context of Baghdad in 2003.

Retired admiral Michael P. Kalleres compared the Tomahawk missile's accuracy of approximately two yards to the accuracy of World War II–era bombs of 300 yards (Fox News Channel March 21). Fox News Channel guest James Carafano also stated that in World War II, 9,000 bombs were required to hit a target; in 2003, only one bomb was necessary (Fox News Channel March 23). Thomas McInerney lauded the speed of the military convoys into Iraq by reminding viewers that it took three months for Allied troops to get from Normandy to Belgium (Fox News Channel March 24). Fox News Channel Pentagon correspondent Major Garrett, like Hunt, justified the coalition's strategy by contrasting it with the strategy of destroying and overtaking cities the Allies used in World War II and United States used in the Korean War (Fox News Channel March 23). When asked to provide commentary for a live firefight in Umm Qasr that seemed to have too many lulls for live television, retired general Wesley Clark remarked that in contrast to the rush of World War II, this firefight was "a matter of a slow, methodical reduction of this enemy position" (CNN March 22). Whether events in 2003 were moving quickly or slowly or proceeding well or not, the references to World War II offered sharp distinctions between the old, imprecise, slow (or fast) way of doing things and the new, satellite-guided, fast (or slow) way of doing things. These comments derived from an unequivocal appreciation for modern military technology and strategy.

Another way that journalists drew upon World War II in their commentary positioned references to the fate of Adolf Hitler as a shorthand for Saddam Hussein's rapidly diminishing power. These references demonized Hussein in order to heighten the threat he posed. Retired navy captain Chuck Nash and retired army major general Robert Scales each relayed the story of Hitler's failed attempt to get his own men to burn Paris in order to illustrate the breakdown in Iraqi command and control (Fox News Channel March 21). Both Scales and panelist Mort Kondracke were concerned that Hussein had a "Hitler bunker complex" and would possibly use WMDs "as a final gesture" (Fox News Channel March 22). Another outright comparison to Hitler occurred when retired colonel Jeff O'Leary stated that Hussein, like Hitler, embedded special security forces with regular troops to guarantee that no one would desert (Fox News Channel March 24). Network journalists also compared Iraqis and Saddam Hussein to another entity that they assumed that most in the United States would see as a historical manifestation of evil and totalitarianism—the former Soviet Union. On CNN, Miles O'Brien described a particular location in Baghdad as Hussein's version of "the Kremlin Square" (CNN March 20). And on Fox News Channel, the Special Republican Guard was compared to the KGB (Fox News Channel March 20). Network journalists did not limit themselves to references to Hitler when they wanted to demonize Hussein, but those references were the most easily communicated. References to World War II offered several quick ways of tapping into commonly understood versions of history to explain and justify the actions of the United States in Iraq.

According to CNN and Fox News journalists, the Vietnam War was an embarrassingly bloody and long conflict that was the antithesis of the 2003 invasion. The Vietnam War played four roles in CNN and Fox News Channel coverage. The first role, which was evident only on CNN, was as another contrasting example that reiterated the humanity of the precision bombings in 2003. Aaron Brown rehashed the B-52 bomber's involvement in carpet bombings in Vietnam, an important comparison to note considering the B-52's revamped job as a precision bomber in 2003 (CNN March 22).

The Vietnam War also represented a loss that the United States had overcome and would not repeat. Here, the pre-written conclusion of the narrative of the 2003 war surfaced. In Hussein's first taped appearance after the invasion began, he made a remark about the lack of stamina of the United States. Aaron Brown interpreted the statement as a reference to Vietnam, and retired general Wesley Clark followed up Brown's remarks by defending U.S. "staying power" (CNN March 20). Clark also described the 2003 army as a "different army" from that of the drafted and bedraggled soldiers in the 1960s and 1970s (CNN March 21). One Fox News Channel correspondent declared that the 2003 invasion was not going to be a Vietnam-like quagmire. Former secretary of state Alexander Haig, appearing as a guest on Fox News Channel, summed up the "remarkably fine job" that the U.S. military was doing by comparing its performance to performance of soldiers in the Vietnam War, when the United States lost 500 soldiers each week (Fox News Channel March 24).

The third role the Vietnam War played in the rhetoric of network journalists was as an affirmation that tragic situations in wartime were normal. Network news personnel sought to reassure viewers that the U.S. military could withstand and overcome the ugliness of war. When a U.S. soldier threw hand grenades into the tents of his superior officers at Camp Pennsylvania in Kuwait, retired colonel Bill Cowan, Chuck Nash, and Rand Corporation adviser/terrorist expert Brian Jenkins relayed stories of similar grenade attacks that both U.S. and Vietnamese soldiers perpetrated on their superiors during the Vietnam War (Fox News Channel March 22 and 23). On CNN, Pentagon correspondent Jamie McIntyre also cited "fragging" incidents in Vietnam (CNN March 22). Likewise, when news broke that Iraqi soldiers were using surrender as a ruse to draw U.S. soldiers into combat, Chuck Nash again noted that these deceptions occurred frequently in Vietnam (Fox News Channel March 24). Further news that previously welcoming Iraqi civilians had turned against U.S. soldiers prompted retired general Don Shepperd to speak of Vietnamese civilians who turned on U.S. soldiers (CNN March 23). Similarly, CNN embedded reporter Walter Rodgers cautioned that coalition forces could never be sure that Iraqi civilians would be friendly and continued that the Mesopotamian delta bore a remarkable similarity to Southeast Asia (CNN March 24).

A fourth way that network journalists used the Vietnam War as an intertextual reference point distanced antiwar demonstrations and war coverage in 2003 from their counterparts in the 1960s and 1970s. On CNN, the references to Vietnam-era protests emphasized that the comportment and message of antiwar demonstrators had improved. Aaron Brown took a low-key approach in his comparison of the protests, describing those in 2003 as "smaller" and "more peaceful" than those of the Vietnam era (CNN March 21). He also noted that one of the legacies of the protests during the Vietnam War was the clarification of the difference between opposition to war and opposition to soldiers (CNN March 20). For Fox News Channel journalists, the anti–Vietnam War demonstrations were a shameful and unpatriotic memory. Geraldo Rivera, who reported from Afghanistan, openly expressed regret for having taken part in the Vietnam protests (Fox News Channel March 22).

CNN and Fox News Channel journalists did not shy away from reflexively speaking of their jobs as journalists in a time of war, and lurking in the background of the coverage was the specter of the Pentagon's linkage of the Vietnam syndrome to media coverage of that war. These references related directly to the improved cooperation evident in the Pentagon-approved embedding system. Former paratrooper John Ringo discussed the post-Vietnam antagonism between the media and the military that Wesley Clark referred to as the "Vietnam media allergy" (Fox News Channel March 21; CNN March 21). Not much discussion was necessary in order to ascertain that the wounds felt by the military still existed. Everyone expressed optimism, however, that the embedded reporters would heal those wounds.

In many ways, the value of the Vietnam War on CNN and Fox News Channel lay in its role as a distancing mechanism, yet in other ways its worth lay in the reminder that the brutality of war could not be exorcised entirely by precision weaponry. The complexity of the Vietnam War broadened the range and meanings of the references, even if network journalists tried to contain those meanings within the rhetoric of progress and patriotism.

References to the 1991 Persian Gulf War, however, sought to recapture an allegedly clear-cut win that was similar to that of World War II. The feeling that the U.S. military "cleaned [the Iraqis'] clocks twelve years ago," as David Hunt phrased it, colored every comment made about the 1991 Persian Gulf War (Fox News Channel March 20). Satisfied that the 1991 war was an example of a just cause and a speedy victory, both CNN and Fox News Channel personnel were able to refer to it in superficially critical ways. References to the 1991 war on CNN and Fox News Channel revolved around the progress since that time that was evident in 2003. By citing the technological and strategic success of the days-old 2003 invasion, CNN and Fox News Channel established the 1991 Persian Gulf War as a slow and technologically inferior antecedent. The repeated references to the 1991 war also mined the false memory of it as a war that had a clear beginning and ending. The U.S. did not stop bombing

Iraq between 1991 and 2003, but that prolongation was distanced from the war proper.[1] Concern that the 2003 war would be a repeat of the long and drawn-out Vietnam War could be seen as unjustified as long as military analysts like David Hunt referred to the 1991 U.S. victory in Iraq as an unequivocal pounding.

References to past wars performed a number of functions that reinforced the idea of the technological and moral superiority of the U.S. military. Network journalists reminded viewers of the good-versus-evil conflict of World War II while comparing that war's destructive bombing campaigns to the cleaner and more precise bombing campaign of 2003. References to the Vietnam War emphasized how far the military had progressed, despite the fact that wounds from that time still had not healed. And the 1991 Persian Gulf War was the source text for the sequel that in 2003 was just beginning to unfold. These references and others used common perceptions of history and heroism to avoid lengthy contextualization, to support the marketable concept of the 2003 war, and to reassure viewers that the 2003 invasion was just, strategically sound, and technologically beyond reproach.

### References to Film Genres

Film genres, the second category of intertextual reference on CNN and Fox News Channel, also served to dispel worries about the 2003 invasion. Network journalists referred to movie genres in an attempt to use formula to contextualize events. The roots, development, and hybridization of film genres and whether genres even exist provide rich material for theoretical debates. But in high concept, genre, like everything else, is a tool of commerce. Genres are presold properties—bastions of predictability that thrive on an audience's engagement with intertextuality. In a broader sense, Hollywood genres are ideological constructions as well as economic ones.

---

1. There were at least four major attacks on Iraq during the Clinton administration. Operation Desert Strike included strikes on fifteen targets on September 3, 1996 (CNN 1994). On June 26, 1993, the U.S. Navy struck the Iraqi Intelligence Agency with twenty-three Tomahawk missiles as a retaliatory response to a plot to assassinate George H. W. Bush (Von Drehle and Smith 1993). During Operation Northern Watch, the U.S. struck Iraq twenty-seven times from January 7 to March 16, 1999 (GlobalSecurity.org. 2005). During Operation Desert Fox, the U.S. bombed Iraq for four days from December 16 to 19, 1998 (Garamone 1998).

High concept's commitment to genre, though inextricably bound to economic imperatives, cannot exist apart from the politics of genre films. Judith Hess Wright asks not what genre films are but what they do. She argues that Hollywood genre films are successful because of their ability to alleviate fears and tension generated by sociopolitical discord. By advocating simplistic and reactionary solutions to the world's problems, genre films uphold the dominant ideology and oppose any disruption to the current order (1995, 41).

Robin Wood argues that genre films are better understood as different manifestations and negotiations of the ideological oppositions inherent in "American capitalist ideology." Ideologically "safe" genre films draw clear distinctions between good and evil, leaving no room for "grays," while others hide the fact of contradiction entirely by "eliminating one of the opposed terms" or by simplifying events (1995, 61–62). The use of opposing agents is not restricted to film. Anne Dunn notes that it is normal for news programs to use binary oppositions, particularly in the "use of conflict to frame stories, and as part of the tendency of news to personalize issues, groups and events as heroes or villains" (2003, 115).

For Wright, simplification or simplistic solutions in genre films represent a socially and politically stagnant worldview that imposes a sense of resignation on those who live in capitalist countries but do not reap the rewards of the capitalist system (43). Though I hesitate to speculate about how genre films affect their viewers, I do see a significant link between the appeal of genre to the system of high concept—the narrative and commercial simplicity of the pre-sold property—and the role that genre films play in upholding mainstream values and consumer culture. High-concept films are not devoid of meaning—they are produced and circulate within social, economic, and political contexts. News programming operates within and endeavors to construct the same contexts.

Throughout CNN's and Fox News Channel's coverage of the first days of the 2003 invasion, references to films and film genres were abundant. References to adventure and science fiction films not only reiterated the binary between good and evil, they also appealed to spectacle and advanced technology to focus on the superficial aspects of war. For example, embedded *New York Times* reporter John Burns described the massive flames of the "shock and awe" bombing campaign as "something out of an *Indiana Jones* movie" (CNN March 21). One U.S. pilot said that "shock and awe" reminded him of *Star Wars* (CNN March 23). These references to movies removed lethal and traumatic bombings from their contexts and consequences.

Both networks referred primarily to war films (particularly World War II films and the film *Black Hawk Down*) and westerns (via the icon of John Wayne, who also represents war films). The references were more than simple allusions to commonly understood texts. In their references to these genres, CNN and Fox News Channel journalists sought to explain and contextualize unfolding events. World War II films depict clear divisions between good and evil and thus are ideologically safe, according to Wood. Judith Hess Wright argues that westerns deal with the contradictory enactments of violence within the texts through the construction of "a code that allows for executions, revenge killings, and killings in defense of one's life and property" (1995, 43). Westerns and war films both promote the dichotomy of good versus evil. In these films, when good and evil are carefully delineated, violence in the service of good is acceptable and even necessary. This attitude toward violence is another reason why these genres are vital as intertextual references for many filmmakers and, as we see, for some newscasters as well.

When CNN and Fox News Channel news personnel referred to genre films, they tapped into cultural knowledge about these films and put that knowledge to work for them. They framed wartime events within marketable formulas and reinforced the line between good and evil. These films represent simplified conflicts and simplistic solutions, and they uphold the status quo. For network journalists, genre was a quick and pre-sold tool of justification.

### References to U.S. Movie Stars

Though references to movie stars were not frequent, they played an important role in establishing and reaffirming ideas about heroism—a noteworthy factor in the representation of the U.S. military and its role in Iraq. Predictably, John Wayne's name made several appearances on Fox News Channel. Whether in a war movie or a western, John Wayne represents the pinnacle of hard-edged and heroic masculinity. Military analyst David Christian referred to Red Adair, the firefighting tycoon of the 1991 Persian Gulf War, as "the John Wayne of putting out wildcat fires" (Fox News Channel March 22). Retired marine pilot Jay Stout dismissed the notion of rescuing POWs by proclaiming, "We don't play John Wayne anymore" (Fox News Channel March 23). In this example, John Wayne represents the maverick hero who gets the job done by flaunting military protocol.

Retired lieutenant colonel Tim Eads drew on the tough-guy persona of Hollywood films when asked what advice the troops needed in order to cope with the news of captured U.S. soldiers. He replied, "What you're gonna tell your troops is, in the famous words of Bruce Willis, 'It's time to cowboy up'" (Fox News Channel March 23). Willis had used a similar line in the film *Tears of the Sun* when his character was about to lead a team of Navy SEALs on a challenging rescue mission. The reference to masculine action-film star Willis was more than apt, given his public support of the war. The BBC reported that when Willis gave a concert for the troops in Iraq and Kuwait in September 2003, he placed a $1 million bounty on Hussein's head and said, "If you catch him, just give me four seconds with Saddam Hussein." In the same BBC report, one military commander described Willis as "a macho actor . . . a guy's guy" (BBC News World Edition 2003).

Conversely, when Jon Scott and Juliet Huddy of *Fox & Friends* referred to "these Hollywood types"—specifically citing Michael Moore and Martin Sheen —the implication was expressly negative. Moore and Sheen were vocal in their opposition to the war (BBC News World Edition 2003; BBC News UK Edition 2003). Scott said he hoped that Moore and Sheen would see the images on television and realize that Iraqis were not "just a bunch of peace-loving desert dwellers," while Huddy said she hoped that they would "support the troops" and keep "their mouths shut" (Fox News Channel March 24). References to movie stars reinforced codes of masculinity and heroism and placed them in the contexts of war and patriotism.

Intertextuality on CNN and Fox News Channel functioned as shorthand to inform coverage of the war. Network journalists tapped into what they believed to be common knowledge to talk about differences between wars, similarities between wars, and justifications for the 2003 invasion. Throughout the 2003 coverage, journalists and commentators at CNN and Fox News Channel drew from a base of knowledge about previous U.S. wars, former antagonists, and media artifacts and stars to steer the coverage in certain directions. CNN and Fox News Channel personnel used intertextuality in a way that gave weight to historical events and to texts or personas that represent those events. Commentators referred to the movie *Star Wars* and a host of movies about World War II to depict wartime experiences that would have been described very differently if told from the point of view of the Iraqi civilians on the receiving end of precision-guided Tomahawk missiles.

### Stars

Stars are constructs that are based on marketability. They minimize financial risk to film studios and ensure that a filmmaker's message will be clearly conveyed. Stars are well-established personalities who rarely stray from the roles they are identified with most clearly. Viewer expectations are essential to the process of casting the star in one kind of narrative or another, and stars and narrative formulas buttress one another. Each star is a way of differentiating the product, yet each is its own standardized product. That dual role allows stars to impart information quickly and easily, and the system that thrives on stars uses them to minimize confusion, limit interpretation, and satisfy viewer expectations. All of this makes the product more profitable.

Media stars function in ways that are similar to how movie and television stars function. Media stars, or newsmakers, typically do not have teams of agents, publicists, and lawyers, but 24-hour news programs repeat information about them for the duration of their newsworthiness and/or marketability. That repetitious bombardment is hardly different from the way in which movie stars make the rounds on television talk shows in the week before their movies premiere. Movie stars and the personas of newsmakers are constructed through repetition and exaltation or degradation. Until stars exhaust their ability to attract an audience to news programming, they remain in heavy rotation. In 2003, CNN and Fox News Channel cast existing stars and created new ones in their war narratives. Each star was an essential element of the marketable concept and an invaluable narrative agent.

### Presidential Stars

CNN and Fox News Channel constructed the two presidential stars, George W. Bush and Saddam Hussein, by repeating information about character traits. The physical absence of both presidents during the first days of the invasion made the task of constructing and reinforcing their personas easier. Bush's absence from public view was striking and unprecedented for a president in wartime (his invisibility was a theme of questions from reporters at White House press briefings in the first days of the invasion), and Hussein was present only through video footage. Because of his Texas accent and less-than-polished public speaking style, Bush has never been an ideal public relations figure. However, in his absence during the first days of the 2003 invasion, on-air personalities on CNN and particularly Fox News Channel spoke enthusiastically in his favor, attempting to alter that negative image. Bush's decision to delay the "shock and awe" bombing campaign was a recurring theme that changed his star persona from the outset of the invasion. The construction of Hussein's star persona dwelled on negativity. Hussein was pre-sold property, and the networks built on his much-publicized persona by dissecting his personality and predicting his fate. Hussein as a concept filled airtime and Hussein as the enemy impeded the Bush administration's political goals. The star personas of these two presidents satisfied the networks' simplistic need for the good/evil binary and established discursive boundaries that avoided complexity.

Just as important as the good/evil binary was the stability that the stardom of the two presidents brought to the war narrative and to the larger political context. Richard Dyer discusses the notion that charisma can assign order to an unstable situation, noting that stars (charismatic beings) can negotiate and reconcile "instabilities, ambiguities and contradictions in the culture" (1998, 31). For example, he points to the "Americanness" of movie star Douglas Fairbanks as a comforting sign for citizens of the United States in the late 1930s. The stardom of Bush and Hussein presents us with a similar situation. Bush's star persona as a strong leader was necessary in order to reassure the U.S. public that a wartime president was indeed in control. And in an atmosphere of fear created by the idea of multiple terrorist cells that could never be entirely eliminated, Hussein represented a single powerful identity onto which a paranoid citizenry could latch. The physical absence of both leaders permitted television journalists and others to creatively construct star personas for both of them that provided a stabilizing narrative in an uncertain situation.

Both CNN and Fox repeatedly circulated the story that the CIA had given Bush credible intelligence about a leadership target that resulted in the postponement of the "shock and awe" bombing in favor of a "decapitation" attack. This narrative enhanced Bush's reputation. CNN's White House correspondents Chris Burns and John King repeated that particular story along with news that Bush had given his generals control over the remainder of the war. King claimed that there was "no sense at all of an agonizing debate" on Bush's part (CNN March 20). Aaron Brown recounted that when Bush gave the order, he said, "Let's go" (ibid.). Brown further commented that if a speechwriter had scripted the moment, the phrase would have been "Let's roll," a reference to the words of Todd Beamer, the passenger aboard United Airlines Flight 93 who led an attack on the terrorists who had hijacked his plane on September 11, 2001, and who is widely regarded as a hero for his actions (CNN March 20).

In an interview with Paula Zahn, former secretary of state Lawrence Eagleburger remarked that he was "impressed" by Bush's decision (CNN March 21). Fox News Channel took a much stronger stance on the subject, cultivating a well-honed persona for Bush as the war president. From March 19 through March 22, Fox News Channel's journalists worked tirelessly to craft Bush's wartime stardom. David Hunt called Bush's decision to "decapitate" Iraqi leadership a "very gutsy call" (Fox News Channel March 19). Retired U.S. navy captain Chuck Nash characterized the attack as a "sign of good, solid leadership" and a mark of Bush's "aggressiveness and ... willingness to seize the opportunity" (Fox News Channel March 20). To Fred Barnes, Bush's order demonstrated that he had an "independent streak" and was "not some guy who just does what the Pentagon tells him to do"; Bush was indeed "acting like a wartime leader" (Fox News Channel March 20 and 21). Brian Kilmeade, who saw the act as an "impressive instance of civilian control of the military," and Steve Doocy, who painted the move as "brilliant" and "brave," praised Bush repeatedly. According to Kilmeade, Bush showed "impressive flexibility," "genuine political control," and intelligence (ibid.). Frequent Fox News Channel guest commentator Bill Kristol was equally effusive, praising Bush's leadership and his dedication to the goal of liberation with minimal casualties. Kristol even declared that Bush "could really change the whole dynamic of the Middle East" (Fox News Channel March 21). Fox News journalists made much of a photograph released by the White House of Bush, Cheney, and Chief of Staff Andrew Card allegedly taken after Bush signed the execute order. Brit Hume called the photo "fascinating," "remarkable," "dramatic," and "compelling" (Fox News Channel March 20). Kilmeade described the photo as "an action shot," which was a bit of an overstatement considering that the shot was of three older men hunched over a desk. In any case, the Fox journalists' interpretation of the entire situation cultivated an image of Bush made almost mythical by the fact that he was rarely in front of a television camera.

On March 19, Bush made a brief appearance to address the start of the invasion. Two days later, Bush made a brief statement to the press with Vice President Dick Cheney looking on. Bush took no questions. That day, Bush left for Camp David, where he spent the weekend. His absence did not go entirely unnoticed. In a White House briefing on that day, one reporter asked White House Press Secretary Ari Fleischer when Bush would address the country next (White House 2003a). On March 24, Fox News Channel's Jon Scott predicted that Bush would be "more visible in the days ahead." CNN and Fox News Channel journalists compensated for Bush's absence with frequent updates and commentary about his last-minute alteration of the war plan, his trip to Camp David, and his day-to-day phone calls, activities, and statements.

CNN and Fox News Channel personnel turned Bush's trip to Camp David into another compelling decision. The first issue centered on the president's departure so soon after the beginning of the invasion. Network commentators perceived Bush's trip as standard because George H. W. Bush had done the same thing in 1991. Shepard Smith viewed the trip as a "good sign," while Jim Angle made it clear that Bush was traveling not for recreation or leisure but to "gather [his] thoughts" (Fox News Channel March 21; CNN March 21). For Fox journalists, the trip was a positive move made by a conscientious and concerned leader.

CNN's John King, Suzanne Malveaux, and Wolf Blitzer repeatedly assured viewers that Bush had the technology at Camp David to stay abreast of the invasion (CNN March 21 and 22). James Rosen asserted that Bush could "rule just fine from there." Neil Cavuto and Molly Henneberg likewise reminded viewers that at Camp David, Bush had many "resources" (Fox News Channel March 22). In making these statements, network journalists reiterated Ari Fleischer's assurance that Camp David "has every modern communication. . . . It has everything that anybody needs," and they sought to provide evidence that Bush was at Camp David to work, not to vacation (White House 2003b).

Network personnel raised another issue that further entrenched Bush's star persona—his ability to carry out his role as commander-in-chief. CNN and Fox News Channel journalists reported frequently on Bush's daily meetings with his war council, and they attempted to ease viewers out of the perception of Bush as the dominant wartime strategist and into an acceptance of his persona as a prudent delegator and multitasker. Both John King and Jim Angle emphasized that Bush was not "micromanaging" the war (CNN March 21; Fox News Channel March 21). Wendell Goler called Bush "confident," and Angle informed viewers that Bush was leaving "the details [of the war] to the military planners" (Fox News Channel March 21). On March 21, 22, and 23, Goler, Cavuto, Henneberg, and Rosen repeated the news that Bush had delegated responsibility to the military experts; Cavuto interpreted this as "a sign of a very comfortable leader" (Fox News Channel March 21–23).

That story, which was repeated seemingly endlessly, coupled with the news that Bush was dividing his time between the invasion and domestic issues, constructed a positive, well-rounded persona. This was a contrast from the persona assigned to the first President Bush during the Gulf War in 1991, when he was criticized for neglecting domestic issues. Reporting on a meeting between the second President Bush and five congressional leaders, Brigitte Quinn clarified that he was "not all Iraq all the time." Additionally, Brian Kilmeade said he was pleasantly "shocked" that Bush had not neglected issues at home in favor of the invasion, a statement that prompted guest Gary Bauer, president of the conservative group American Values, to call Bush "a smart president" (Fox News Channel March 21).

From March 19 to March 23, CNN and Fox News Channel news personnel painstakingly constructed a favorable image of President Bush. The White House prompted the networks by supplying photos and statements, but the frequency of the positive reinforcement provided by the reporting and editorializing at the two networks did more for Bush's persona than any White House press conference could have. Network journalists saturated the first two days of the invasion with discussions of Bush's order to strike Iraqi leadership targets and thus alter the war plan. In their version of the story, Bush was an active, decisive, and bold leader. His trip to Camp David threatened to compromise his persona as an active leader, so CNN and Fox News Channel emphasized his ability to maintain his work schedule from his vacation site. Bush's "decapitation" attack order was a one-time occurrence, so CNN and Fox News Channel underscored his wisdom in stepping aside to let the experts handle the day-to-day operations. They solidified their story of his astuteness by emphasizing that he continued to work on domestic issues during the invasion of Iraq. By adhering to a limited set of topics released by the White House and by repeating those topics with frequent, upbeat editorializing, CNN and especially Fox News Channel took on the role of publicists, creating a star out of a key player with a theretofore less-than-stellar image.

The two networks performed just as diligently with Bush's Iraqi counterpart. The first Bush administration had campaigned aggressively to make the Hussein name synonymous with evil, and the effects remained. CNN's Judy Woodruff stated that the second Bush administration portrayed Hussein as a "ruthless dictator" and the "embodiment of evil" (CNN March 24). The characterization of Hussein as a pathological murderer with paranoid tendencies formed a key part of his public image and a key part of the marketing of the 2003 invasion. A less-concerted public relations effort or a policy of letting Hussein's deeds speak for themselves might have fallen below the radar of the U.S. public. But with the focused campaign of two White Houses and CNN and Fox News media personnel, the former Iraqi president became an infamous star but a star nonetheless. That image or star persona, created over more than a decade, was the product of his violent regime, his previous (and ongoing) conflicts with the United States, and the second Bush administration's need for a dire threat. Despite the dearth of evidence connecting Iraq to the terrorist attacks on the United States in 2001 and despite the dubious nature of the claim that Iraq had a stockpile of WMDs, Hussein had a reputation that was tarnished enough that Bush was able to use it to galvanize public and congressional support for war. The launch and success of the 2003 invasion of Iraq depended on the administration's ability to create fear among U.S. citizens, or, as Giroux would argue, to construct and implement the spectacle of terrorism.

The construction of Hussein on 24-hour news networks followed the same blueprint. Both CNN and Fox News Channel journalists emphasized Hussein's long-established persona as an evil villain in their coverage of the war. On Fox News Channel, Hussein was the "Butcher of Baghdad" and part of a "subhuman species" (Fox News Channel March 22 and 24). Hussein's image and name inevitably invoked his past as the invader of Kuwait. As further evidence of his infamy, CNN and Fox News Channel personnel referred to crimes against humanity of which he was guilty.

Hussein was the face of the antagonist in the war, so his fate was understandably of great consequence to the war narrative. From March 20 through March 24, CNN and Fox News Channel were abuzz with speculation about whether Hussein had been killed in the "decapitation" strikes on the Iraqi leadership on March 19. Senator John McCain revealed that he had talked to people who did not believe that Hussein had been killed, Pentagon correspondent Barbara Starr conveyed the Department of Defense's uncertainty about Hussein's demise, and Nic Robertson reported the claim by Iraqi officials that Hussein was indeed alive (CNN March 20).

The question of whether Hussein was dead or alive was a mainstay topic punctuated by videotapes of Hussein in meetings or giving statements. On March 20, the Iraq Information Ministry released a video recording of Hussein reading what appeared to be a handwritten statement. CNN and Fox News Channel journalists focused on the authenticity of the video and what it revealed about Hussein's fate. The video was the first of many images of Hussein that would spark debates about when the images were recorded. This first recording and the resulting flurry of supposition, analysis, and more supposition reinforced the notion that Hussein—the reason behind the war—was a construction. The person on the video might not have been him, or at least that was the first reaction of network journalists. Tales of body doubles, a suspicious-looking birthmark, alarmingly oversized glasses, a puffy face, and a mistress/informant ran amok until March 21, when both networks announced that the CIA had confirmed the man in the recording was indeed Hussein. To an extent, whether the recording was of Hussein was immaterial. The mere idea of Hussein prolonged the discussion of the threat he posed. Moreover, the figure of Hussein filled screen time, created anticipation, and detracted from other more pressing issues. Each subsequent video recording of Hussein the Iraqi government released triggered the same search for legitimacy and raised exactly the same questions.

On March 24, Aaron Brown asked Nic Robertson, who was reporting from Jordan, if anyone there was discussing Hussein's status. Robertson replied that the issue was not being discussed in quite the same manner as in the United States and Britain (CNN March 24). His point was salient, given that the ubiquitous question—dead or alive?—hardly mattered and was fueled only by guesses and hearsay.

Because the administration had branded Hussein the face of the enemy, his fate was a frequent topic on CNN and an absolute obsession on Fox News

Channel. From the beginning of March 21 to the end of March 22, and in the context of news stories rather than updates, Fox News Channel broached the topic of Hussein's possible death at least eighty-nine times.[2] CNN did so at least twenty-eight times. On both CNN and Fox News Channel, journalists frantically posed the question about Hussein's fate to each other and sometimes to guests who rarely had any way of knowing the answer. For example, CNN's Paula Zahn asked the question of Senator Tom Daschle and Arab League ambassador to the UN Yahya Mahmassani to no avail (CNN March 21). On Fox News Channel, Rita Cosby asked Bill Cowan, terrorism expert Matt Epstein, and Henry Kissinger (who said he did have information that Hussein was alive); E. D. Hill asked *Weekly Standard* editor Bill Kristol; and John Gibson asked Andrew Cockburn, coauthor of *Out of the Ashes: The Resurrection of Saddam Hussein* (Fox News Channel March 21 and 22). Fox News Channel journalists characterized the incessant query as a "very heated discussion" and a "great debate," but the excessive questioning only stimulated what was little more than conjecture (Fox News Channel March 21 and 22). One story pertaining to Hussein's fate that Fox News Channel ran (but CNN did not) was attributed to "U.S. officials" who said they had "images of panicked digging" at a bombed bunker (Fox News Channel March 21). The images also allegedly showed Hussein being loaded into an ambulance on a stretcher. Fox News Channel aired this story no less than thirty-eight times from March 21 through March 24.

The issue of whether or not Iraqi leadership was intact appeared much less often than the tantalizing "dead or alive" question, but it served as a bridge to the more personal discussions of Hussein's emotional and psychological states. In a Department of Defense news briefing on March 22, Victoria Clarke quickly mentioned a state of "confusion" in Iraqi command and control (U.S. Department of Defense 2003e). Clarke repeated that claim on March 24, reporting that the U.S. military continued to see "signs of some confusion and disarray" among the Iraqi leadership (U.S. Department of Defense 2003f). Like the Department of Defense, CNN and Fox News Channel maintained a consensus about the

---

2. I selected these dates because my videotaped coverage for those two days contained the fewest gaps in time. I am missing seven minutes from CNN footage and one hour and eighteen minutes from Fox News Channel footage for March 21. I am missing three minutes from CNN and forty-five minutes from Fox News Channel for March 22. Even the absence of just over two hours from Fox News Channel footage skews my results so that my count is substantially lopsided.

disintegration of Iraqi control, but even that story gave way to another attempt at personalization. Conjecture about Hussein's physical status was subsequently augmented by speculation about his psychological state.

Among those on CNN who allegedly spoke with authority about Hussein's indisputable state of paranoia were Ambassador Joe Wilson; Kenneth Pollack, author of *The Threatening Storm: The Case for Invading Iraq*; and Simon Henderson, a Hussein biographer. On Fox News Channel, Harlan Ullman, co-creator of the "rapid dominance" war strategy (also known as "shock and awe"), called Hussein "incompetent" and claimed that he was "in denial" (Fox News Channel March 21). In addition to assessing Hussein's paranoia, former CIA operative Wayne Simmons called him a "homicidal maniac." Andrew Cockburn characterized the Iraqi president as a "tough guy from the streets" and a "gambler" who should be used to this type of situation (Fox News Channel March 21). Military analysts David Christian and Robert Bevelacqua both unequivocally agreed that Hussein was "very paranoid" about the possibility that a mole was in his organization (Fox News Channel March 23). The amateur psychological analysis of Hussein created more drama in the networks' representations of his life. The networks did not know if he was dead or alive, but if he was, the narrative insisted, his ability to command was being hampered by paranoia about possible traitors. The drama became a psychological thriller, a twist that gave credibility to yet another offshoot of Hussein's star persona: his proclivity for violence.

Hussein's violent behavior was a topic that helped sustain his importance to the war narrative and therefore his stardom. At the outset of the invasion, David Ensor, CNN's national security correspondent, reiterated one of the official justifications for going to war: Hussein "may order the use of the very WMDs he insists he does not have" (CNN March 19). Ensor's statement was preceded by a file photo of Hussein holding a sword—an image that compounded the threat Ensor voiced. The risk that Hussein would unleash his WMDs was a talking point the Department of Defense had clearly established. On March 20, the day after the war's opening strikes, Rumsfeld reminded the press that Hussein could order the use of WMDs on anyone at any time (U.S. Department of Defense 2003c). On Fox News Channel, Laurie Mylroie, author of *Saddam Hussein and the World Trade Center Attacks: A Story of Revenge,* discussed Hussein's "willingness to use . . . incredible brutality" (Fox News Channel March 20). Fred Barnes underscored the stated aims of the U.S. military to minimize civilian casualties by stating that Hussein wanted just the opposite, and Amy Kellogg speculated that Hussein might "lash out at Israel" if he felt "boxed into a corner" (Fox News Channel March 21 and 22). As long as Hussein was alive, his penchant for violence continued to instill a sense of urgency in the marketable concept and the networks' narrative of the war.

As the war progressed, Hussein's persona remained front and center for some but not for others. Early in the invasion, Paula Zahn asked retired colonel Mike Turner if the campaign would be a failure without the capture of Hussein. Turner replied that President Bush had made it clear that Hussein needed to be deposed and that the capture of Hussein was necessary for a victory. However, later in the day, Wesley Clark warned against becoming "fixated on the personage of Saddam Hussein" (CNN March 20). For Clark, Hussein was important because he was the head of command, not because of his persona. Two days later, CNN's White House reporter, Suzanne Malveaux, said, "It's not about one man or one personality," reminding viewers that the administration's attitude toward Hussein paralleled its attitude toward Osama bin Laden (CNN March 22). Yet there was tension between Bush's need for good public relations ("It's not about one man") and the need of those who promoted the war narrative of personalized icons. Both Hussein and bin Laden were the recognizable faces of the enemy in the war on terror. Yet the statement the White House put out called Hussein's stardom into question.

CNN's Aaron Brown spun the tension between public relations and the war narrative by noting that the U.S. public was the consumer of the personas the networks (and the Bush administration) had constructed: "Given the way that Americans process information . . . and tend to personalize these sorts of conflicts, I'll just gently suggest that . . . for domestic political reasons, he's got to be found, one way or another" (CNN March 22). Brown blamed U.S. citizens for personalizing the war and did not assign any culpability to the way in which CNN had repeated the Bush administration's personalization of the conflict. Wesley Clark delivered a more insightful comment than Brown:

> Somehow, a lot of people believed that Saddam was responsible for 9-11, and the administration has worked very hard to prove linkages between Saddam and Osama bin Laden, and to some extent we've almost transposed our anger at Osama bin Laden and fixated on Saddam, and Saddam is a bad guy. But if we get rid of his control over that state, Saddam doesn't matter. (ibid.)

Clark, who was a military news analyst, not a network journalist, came dangerously close to conceding that Saddam Hussein was not connected to the 9/11 terrorist attacks on the United States. This obviously would have been a departure from the official narrative of the war, which relied on the U.S. public's fear of another terrorist attack. Clark's statement also placed blame for the Hussein fixation on the administration. Though he agreed with Clark, Ken Pollack stated that anything less than the capture of Hussein would be an "embarrassment" for the administration (ibid.).

The debate about Hussein's use value was similar on Fox News Channel. Bill Cowan adhered to the theme of regime change when anchor John Gibson asked him if the point of the war was personal (Fox News Channel March 20). When Rita Cosby asked Cowan if he thought Hussein was alive, Cowan rebuffed her question, stating that it did not matter because the war as a whole was a more pressing issue. Robert Scales offered contradictory viewpoints on the subject. First he cautioned against "focusing too much on the man" at the expense of concentrating on the mission (Fox News Channel March 21). In another conversation, he again warned against focusing on the individual instead of focusing on the regime, to which Greta Van Susteren replied, "But he is the regime." At the conclusion of their conversation, Scales retreated, claiming, "If Saddam is dead, the campaign is pretty much over" (Fox News Channel March 22). The lack of consensus on Fox News Channel did not detract from the enduring value of Hussein's persona to the marketable concept and the narrative of the war.

*Military Stars*

The U.S. military occupied a cherished place in the coverage of CNN and Fox News Channel, and the faces of high-ranking military personnel became stars in their own right at the two networks. The face of Operation Iraqi Freedom was General Tommy Franks, the allied commander. However, Franks's demeanor was controlled and subdued. He did not play well as a media star. The real military star, the telegenic figure who catapulted to visibility, was Iraqi information minister Mohammed Said al-Sahhaf. With his commitment to shameless propaganda, Mohammed Said al-Sahhaf put on the show that Franks did not. The networks needed his presence to support their war narrative, and they built his persona to suit their purposes.

Franks' first appearance was highly anticipated by an array of reporters assigned to the media center at CENTCOM in Qatar. The set from which Franks delivered his briefings had been designed by Hollywood art director George Allison and cost $1.5 million to build. March 20 and 21 passed with frequent reports from CNN and Fox News Channel's CENTCOM-based reporters about the "news vacuum" and "news blackout" at the big-budget media center (Fox News Channel March 20 and 21). Frustrated by the briefing delay, Rick Folbaum complained about the "fancy schmancy war set" that he felt was built for apparently no reason (Fox News Channel March 22). Brit Hume described CENTCOM's briefing stage as a "Hollywood set," an accurate description, and

Fox's David Lee Miller characterized the much-anticipated upcoming briefing at CENTCOM as "something akin to a premiere" (Fox News Channel March 21 and 22).

After Franks gave his long-awaited first briefing, Miller referred to it as "the General Tommy Franks Show" (Fox News Channel March 22). Journalists at both networks noted Franks's low-key approach, contrasting it with General Norman Schwarzkopf's more animated briefing style. Franks was a serious, no-nonsense military man, a strong and silent type. Because he was a member of the U.S. military, he belonged to a group that CNN and Fox News Channel regularly praised, but beyond that, Franks was wallpaper.

Mohammed Said al-Sahhaf could not have offered a stronger contrast. The Iraqi information minister was the most visible face of the Iraqi regime during the first days of the invasion. U.S. network journalists, ignorant of who was who in the Iraqi regime, could not even correctly identify him at first; Fox's Shepard Smith mistakenly identified him as "an Iraqi television news announcer" on March 20. But soon he was a well-known figure to the U.S. broadcasters, and they quickly began constructing an image of him as untrustworthy and perhaps even mentally ill. Fox's Laurie Dhue labeled him a "delusional" propagandist, for example (Fox News Channel March 22). Each press conference the information minister gave generated strong criticism from Fox News Channel news personnel. On March 21, Lauren Green said, "They're clearly stating their point of view and their side of the story, wanting to gather world support for themselves and against the U.S." A short time later, Molly Henneberg asked sarcastically, "What's this I'm hearing about misinformation coming from the Iraqi Information Ministry? Are you serious?" (Fox News Channel March 21). E. D. Hill called him a liar, Brian Wilson observed that he was "giving a very different picture" of the invasion, and David Christian accused him of spreading propaganda (Fox News Channel March 21 and 22). The criticisms and accusations that Fox News Channel journalists leveled at the information minister had merit; no U.S. viewer watching the embedded coverage and buying into it would dispute the notion that his press conferences were full of distortions. But Fox News Channel so battered his authority that his briefings became a sideshow of which he was undoubtedly the star. He fulfilled Fox News Channel's expectations for an excessive propagandist, and they constructed his persona out of their incredulity. He was not a star on the scale of Hussein, but he was an ersatz figurehead that appeared to viewers on live television when the actual figurehead could only be found on video recordings.

Military personnel occupied a subordinate space in the war coverage in relation to the presidential stars, but in the logic of the war narrative, they were necessary icons representing the two sides of the invasion. Like Bush's absence, Franks's infrequent briefings made it necessary for network journalists to build his persona through discourse. And like Hussein's attention-grabbing appearances on videotape, al-Sahhaf's wildly inaccurate recorded statements incited a great deal of discussion and debate. In the personas the network journalists constructed for the two men, Franks was an icon of the no-nonsense U.S. military of the marketable concept and the flamboyant al-Sahhaf was a discreditable figure who further marginalized the already-derided Iraqi military.

*In-House Stars*

The reporters embedded with the U.S. military were perhaps the biggest stars of the war at CNN and Fox News Channel. They were second only to the presidential stars of the conflict. They became media stars during the war for two reasons: their reports were exciting and telegenic and their status as journalists was controversial because of their high degree of cooperation with the military. Indeed, CNN and Fox embedded journalists were eager purveyors of information the military fed to them.

In their role as media stars, embedded reporters participated in disseminating military propaganda. Whether it originated from the Iraqis or the coalition, military-based propaganda was integral to the operation of the war, according to Robert Scales, a military analyst for Fox News Channel and NPR (Fox News Channel March 21). For Scales, the war was about information, particularly whose information could do the most psychological damage to which side's troops. Embedded reporters were key players in this larger strategy, so they became key faces in the coverage.

These journalists functioned as stars in multiple ways, propping up the goals of the U.S. military while putting on a show for all who wanted to watch. The appeal of embedded reporters to their respective news organizations (and their primary selling point in the coverage of the war) was the element of realism in their reports. As with all mediated representations, reports of embedded journalists contained elements of illusion masquerading as objective truth. Fiske (1987) argues that the claim to objectivity is an ideological mechanism; in the case of reports from embedded journalists, all of the touches of realism in the reports they transmitted—the grainy visuals, sonic disconnects, unkempt reporters—obscured the "constructedness" of the reports and the ideology enveloped within them. For Fiske, "Grounding ideology in reality is a way of making it appear unchallengeable and unchangeable, and thus is a reactionary political strategy" (36). The "reality" of embedded reporters was pure spectacle, and this was precisely what the U.S. military hoped to achieved with the new cooperative system. Embedded reporter Greg Kelly summed up the illusion of realism when he said that the embedding process was "almost like a media event. When the press shows up at the appointed place . . . we're shown what we came to see" (Fox News Channel March 20). By pointing a camera at what the military showed them, embedded reporters chronicled the war narrative and publicized the marketable concept.

CNN and Fox News studio journalists made sure that viewers understood the star value of their embedded reporters. In his introduction to one segment with embedded journalist Greg Kelly, Shepard Smith announced the growing appeal of such reporters within the news organization and described Kelly as "one of our newest who made a name for himself" (Fox News Channel March 21). Most of the star reporters who were embedded were men, and all of them spoke of the exploits of their units with such fervor that the energy was palpable, even over a satellite phone.[3] Walter Rodgers exclaimed that hostile Iraqi fire "put steel into the spine of the American soldiers." He also excitedly referred to the military convoy as a "wave of steel" (CNN March 20). Rick Leventhal described the British and U.S. Marines as "gung ho," adding that the marines were "doing an outstanding job" and were "always ready to roll" (ibid.). Although embedded reporters were not fighting the war themselves (some were former soldiers, actually), they seemed to relish the job of hangers-on, as if proximity to the soldiers boosted their participation in the events. To this end, one curious slippage came on March 20 when the Fox News Channel banner read, "Fox troops with U.S. troops heading into Iraq." The text appeared only once, after which the wording was altered to read, "Fox news crew with U.S. troops in Iraq" (ibid.). The conflation of embedded reporters with the troops clearly encapsulated Fox News Channel's attitudes toward their reporters.

Their emulation of the military made embedded reporters highly sought after by anchors in the network studios. The novelty of a live embedded report

---

3. CNN's most visible embedded journalists in the first five days of the invasion were Walter Rodgers (3rd Squadron, 7th Cavalry, 3rd Infantry Division), Martin Savidge (1st Battalion, 7th Marines, 1st Marine Division), Ryan Chilcote (101st Airborne Division), Dr. Sanjay Gupta (1st Medical Battalion), and Jason Bellini (15th Marine Expeditionary Unit). Fox News Channel's most visible embedded reporters were Rick Leventhal (1st Marine, 3rd Light Armed Reconnaissance), Greg Kelly (2nd Brigade, 3rd Infantry Division), Kevin Monahan (USS *Constellation*), and Steve Centanni (Navy SEALs).

in the first few days of the invasion was evident; on-air personalities would interrupt studio interviews and even cut sentences short to switch to incoming reports. Whether the reporters were dressed in desert attire (as in the cases of Walter Rodgers and Rick Leventhal) or standing on the decks of an aircraft carrier lit by an eerie red glow, they represented the action and danger war correspondents faced. Much more than the U.S. soldiers, embedded reporters were the stars of the invasion because they spanned two worlds—the battlefield and the news studio. They were marketed as being able to speak with authority about both arenas and able to import context and perspective about the war zone as they provided action-filled details about what was happening overseas. Not surprisingly, the product fell short of the marketing, but the result was no less spectacular.

Part of the spectacle of embedded reports derived from the logistics and politics behind the embedding system. The controversy that it ignited was a frequent topic on CNN and Fox News Channel. Those self-referential moments in which on-air personalities discussed the embedding system not only justified the system and reaffirmed its health in spite of critics' warnings, but also added another dimension of "it-ness" to embedded reporters who were already enjoying a great deal of attention from the military, media critics, and media practitioners.

The "it" factor was problematic, however, and post-embedding analyses have yielded mixed attitudes toward the reporters' roles in the war effort. Although some military leaders complained about incidents when reporters disclosed sensitive information and on-air military analysts second-guessed the war plan, they generally saw the results of embedding as positive (Purdum and Rutenberg 2003; McLane 2004, 83).[4] These leaders cited "trust," "confidence," "personal relationships," and "understanding" as key to the military's strategy

---

4. At a workshop entitled "Reporters on the Ground: The Military and the Media's Joint Experience During Operation Iraqi Freedom" led by the United States Army War College's Center for Strategic Leadership, military personnel readily described how they "used the media" to win the information war (Pasquarett 2003). For the military, embedding was a way to put "a face on a battle" and to tell the story of the U.S. soldier to the U.S. public (Miracle 2003, 45; Kosterman 2003). In fact, the military viewed journalistic objectivity as "irrelevant"; only trust between embedded journalists and soldiers could ensure the "integrity" of reporting (Pasquarett 2003). In this way, embedded journalists no longer worked for their respective news organizations; they worked for the U.S. military. Consequently, the military was mostly pleased with the results.

of making embedded reporters their de facto employees (Miracle 2003, 42; Pasquarett 2003; McLane 2004, 83; Kosterman 2003, 3).

Trust, confidence, and understanding were discernible in the relationships between embedded reporters and soldiers and in the adherence of reporters and news organizations to the U.S. military's embedding rules. These rules upheld what the military called "operational security." Embedded reporters were instructed to protect the secrecy of the missions, and in doing so, they stepped further into military terrain and away from their day jobs. They respected operational security for the most part and plainly revealed when their military minders prohibited them from speaking about missions altogether. The studio-bound journalists in the United States also pledged to adhere to the rules. For example, on March 19, when embedded reporters could not yet make contact with the studios, Shepard Smith stood by the military by not pressing the issue. He simply said, "We'll leave it at that" (Fox News Channel March 19). Pentagon correspondent Bret Baier also took a certain amount of pride in following the rules when he said that Fox News Channel released only what "senior defense officials have said that we can" (Fox News Channel March 20). At CNN, Aaron Brown commented that the embedding process "works because we all understand the rules and . . . the consequences." Throughout, Brown praised the embedded journalists and swore that they were not revealing more than was permitted (CNN March 20). Carol Costello likewise assured viewers that embedded reporter Walter Rodgers would never put troops in any danger, and military analyst Wesley Clark said that the embedding rules provided parameters that ensured that everyone was careful (CNN March 21 and 22).

While the studio-based network journalists remained separate from the battlefield and from operational information (until senior defense officials cleared it), embedded reporters had privileged access to information. That was the understanding, at least. CNN and Fox News Channel news personnel insisted that the embedding system created and maintained "openness" and "unbelievable access" (Fox News Channel March 20; CNN March 22). However, the embedded journalists were not as privileged as the U.S. military or CNN and Fox journalists would have had us believe. National Public Radio reporter John Burnett, who wrote a self-critical analysis after his embedding experience, noted that that he and his colleagues "lived exclusively within the reality of the U.S. military," adding that "the inability of embedded journalists to verify the military's version of the war in Iraq made for one-sided reporting" (2003). Nevertheless, CNN and Fox News Channel news personnel praised the rules, the system, and the Pentagon.

By extolling the virtues of the entire embedding package, CNN and Fox News Channel declared an end to the historically tense media-military relationship. Not only was the embedding "historic television journalism," it was also "extraordinary," "amazing," "phenomenal," "fascinating," "dramatic," and according to Bill Hemmer, the "single greatest source of information" (CNN March 20 and 21). Whereas Christiane Amanpour contextualized the embedding system as the result of frustration with the "draconian censorship of the [1991] Persian Gulf War and Afghanistan," Fox News Channel personnel focused on the present (CNN March 20). Shepard Smith characterized embedding as the Pentagon's "amazing endeavor" and applauded the cooperation between the media and the Department of Defense (Fox News Channel March 20). Bill Cowan plainly said, "I give a lot of credit to the Pentagon for allowing this to happen"; and Steve Doocy called the Pentagon "brilliant" (Fox News Channel March 21, 23). Fox News Channel praised the new system as a "really remarkable level of openness . . . that we've never before seen in human history" (Fox News Channel March 21). The idea that the system worked elevated embedded reporters to the status of journalistic heroes who were blazing trails and making peace with the military; in short, these reporters were healing the wounds from the Vietnam War.

With that degree of support from CNN and Fox News Channel, it is easy to see how embedded reporters could be perceived as the Pentagon's stars, too. Colin Soloway, a *Newsweek* embedded reporter, remarked how "useful" it was to see "things from the military's perspective" (CNN March 21). Fox News Pentagon correspondent Major Garrett pointed out that the reporters were "scouts, not only for [the news media] . . . but scouts for the Pentagon" (Fox News Channel March 20). Later, as Rita Cosby and Gregg Jarrett spoke optimistically about the historic cooperation between the media and the military, Garrett interjected that the system was "beneficial to the Pentagon as well" (Fox News Channel March 21). In an article for *Military Review,* Tammy L. Miracle writes, "If the public believes embedding journalists is a way for the Pentagon to control the news rather than to report it, the Army will have gained nothing" (2003, 44). Miracle's choice of words is telling; the success of the embedding system depended not on whether the Pentagon was actually controlling the news but on how the public interpreted the Pentagon's use of embedded reporters.

CNN and Fox network journalists made no secret about the Pentagon's desire for controlled, positive images from the battlefield. Paula Zahn overtly stated that the images beamed in from embedded reporters were "the images the Pentagon had been hoping for" (CNN March 22). Guest commentator Jane

Perlez of the *New York Times* commented that "real-time imagery" from embedded reporters was "good for the Pentagon" (CNN March 24). Pondering the embedded footage of the U.S. military's march into Baghdad, Shepard Smith speculated, "Surely that must have been part of the Pentagon's planning when they figured the best way to get the message to Saddam Hussein is to show it to him on at least ten satellite channels and networks around the globe" (Fox News Channel March 21). Later, Smith asked Pentagon correspondent Bret Baier if the system had "paid off in the way [the Pentagon] clearly hoped it might," and Baier replied that the Pentagon was "very happy with these images so far" (Fox News Channel March 22). These examples dispute Brit Hume's claim that embedding did not mean "in bed with" (Fox News Channel March 21). Although Hume apparently did not anticipate any breach of journalistic objectivity—an elusive idea in and of itself—objectivity was an issue for some embedded reporters. Despite the relative absence of direct censorship, former embedded journalist John Burnett concluded that embedding "served the purposes of the military more than it served the cause of balanced journalism" (2003).

Burnett's experiences of not being able to see the effects of artillery and of not being able to "verify the military's version of the war" reflect the concern that the reports of embedded journalists sacrificed context for immediacy (ibid.). In the March 21 Department of Defense briefing, Rumsfeld admitted to the press,

> What we are seeing is not the war in Iraq. What we're seeing are slices of the war in Iraq. We're seeing that particularized perspective that that reporter, or that commentator or that television camera happens to be able to see at that moment. And it's not what's taking place. What you see is taking place, to be sure, but it is one slice. (U.S. Department of Defense 2003d)

The Pentagon also described the lack of context as looking through a "soda straw." Former White House situation director Michael Bonn elaborated on that description by saying, "All the soda straws feed into the situation room" (Fox News Channel March 24). These metaphors implied that to some extent, lack of context did not matter. What mattered was that the military—not television news or television viewers—had the "complete" picture.

As with many concerns about embedding, the U.S. military and media critics viewed the lack of context on television from different perspectives. In his post-invasion analysis of the embedding system, Lieutenant Commander Brendan McLane expressed concern about decontextualized reporting, but he had a very specific definition of context. For him, a lack of context meant a lack of comprehension of the tactical and operational objectives of the war (McLane

2004, 86). He considered embedded journalists to be employees of the military and not employees of news organizations that should report the details of military engagement within a larger political context.

An incident on March 23 illustrates McLane's priorities and concerns. At approximately 9 AM Baghdad time, during what appeared to be a major firefight at Umm Qasr (but what was actually "nickel-and-dime stuff," according to retired colonel Ed Badolato), military analyst David Christian and other military analysts at Fox News Channel were concerned that the news camera recording the firefight was pointed in the wrong direction, that the soldiers were playing to the camera, and that the British reporter with the coalition troops had little knowledge of the military tactics in operation. Military analyst David Hunt recalled that embedded reporter Rick Leventhal had asked a marine to recount the events of a firefight, but at Umm Qasr, Hunt remarked that the "Brit" sounded like he was "talking about a soccer match" (Fox News Channel March 23). Aside from the obvious preference for U.S. journalists in this instance, military analysts placed pressure on the reporters to demonstrate knowledge about all things military to contextualize events from the military's point of view.

As spokespeople for the military and as stars, embedded journalists were successful. Their adventurous reporting contrasted with the absence of similar access in the 1991 Persian Gulf War. Their unfettered praise for the troops ("You're doing a heck of a job out there," said Leventhal to a soldier) and their open assistance to the Pentagon solidified them as true patriots in a war that was presumably being fought to avenge and defend the U.S. way of life (Fox News Channel March 21). Particularly on Fox News Channel, any criticisms leveled at the U.S. military's use of journalists were seen as irrelevant. The main issues were that the rules were being followed, the system worked, the Pentagon was the hero for brokering an unprecedented deal between the military and the media, and the news networks could help combat Iraqi propaganda by telling the U.S. military's story. That was the important story, and the embedded reporters, even more than the retired military officers serving as analysts, told that story with fervor.

The story did not create the stars, however; it was the repetitive and positive commentary of network news personnel that did that. The extensive preinvasion logistics the military created—which included laying down ground rules, boot camp for embedded journalists, and deployment—coupled with the discourse surrounding the embedding system generated terrific buzz. CNN and Fox News Channel journalists sustained the buzz, especially in the first five days of the invasion, and in doing so they ensured that embedded journalists would achieve and retain status as chroniclers of the revenge narrative and publicists of the marketable concept.

## Conclusion

High-concept films depend on intertextuality to perform two major functions. Intertextual references, whether they operate using genres or stars, mine the cultural knowledge of viewers as a means of transmitting narrative information. This type of shorthand conserves screen time, but it also functions expediently in the marketing arena. Limited advertising space on television or in publications can be used most efficiently with symbols and icons that refer to other well-known texts in a way that succinctly enriches the context. Although 24-hour news programming has a great deal of time to offer contextualized information, I did not find such contextualization in CNN and Fox News Channel's war coverage.

Instead, I found a wealth of intertextual references that replaced wide-ranging explanations of the invasion of Iraq. CNN and Fox News Channel studio journalists invoked specific memories of past U.S. wars to explain and justify the trajectory of the current one. The references on the two networks to genres contained the war's events within pre-sold formulas that reconciled contradictory ideas about violence in the United States by creating a code of acceptable violence. Furthermore, network news personnel built star personas and recycled established ones, acting as the journalistic equivalent of publicists while relying on the cult of stardom to drive the narrative forward and calm instabilities.

# 4

## *War Characters*

On March 22, Geraldo Rivera, one of Fox News Channel's maverick reporters who was stationed in Afghanistan, used a bit of well-worn rhetoric to imply that antiwar protesters were somehow acting in concert with Middle Eastern extremists. He directed viewers' attention to a map of the Middle East and said,

> Look at that band, that belt of absolute threat to freedom-loving people everywhere. You know, I just wish that the antiwar demonstrators would pull out their atlas one day and . . . look at this potential anarchy, a belt where governments heretofore have been extremely hostile to anything that democratic-loving people believe in. (Fox News Channel March 22)

Rivera's suggestion that protesters at home were a threat to democracy was the beginning of an extended effort to characterize antiwar protesters as traitors to the U.S. government.

Like stars, character types have a function in terms of the narrative, in terms of the marketplace, and in terms of ideology. The reductive nature of a type is of great intertextual value because it draws on viewers' knowledge of other embodiments of that type. Types also have exchange value: they simplify the process of creating characters, they fulfill audience expectations, and they are easy to market because their motivations are easy to communicate. Finally, types blunt complexities. Types are easy to love and hate because they do not allow room for gray areas; as a result, they are ideal components of a product's marketable concept. This chapter traces the use of the basic character types of heroes, villains, and sidekicks at CNN and Fox News Channel. A close reading of the networks' character typing also reveals an unexpected and decidedly more complex character known as the false hero, a type Vladimir Propp elucidated. The networks' character types were possibly the most blatantly constructed elements of the war narrative because network news personnel were able to control the process of constructing them.

## The Value of Types

Sarah Kozloff argues that television's formulaic structure invites us to look for rules and configurations that form the basis of television programs (1992, 72). Character types help in this process, and though the characters on television programs are quite formulaic (typified by such well-worn tropes as detective or father or working woman), Kozloff argues that "the viewer's interest is continually engaged by the personalities who fulfill these roles" (76). Not all television characters are as cleanly drawn as Kozloff proclaims—the amount of character development engendered by the serial narrative is one strong argument against Kozloff—but one cannot deny that formulaic characters are essential to commercial television.

Formulaic and predictable character types are narrative and ideological tools that consistently attract audiences and endow events with loaded meanings. Karim H. Karim discusses the implications of character typing in violent storylines:

> Actual physical violence is endowed with high symbolic value when it is depicted as supporting one or another view of political and social reality. Narratives that dramatize deadly struggles involving heroes, villains, and victims give meaning to the conflicts that exist in human society. Portrayals of particular uses of violence by specific kinds of people identify which social roles they are playing. Those who control the production and dissemination of these dramaturges have the means to influence public opinion regarding the types of people that are to be considered as the heroes, the villains, and the victims in society. (2003, 19)

Karim's argument forces us to consider the effect that character typing has on actual participants in a war. As I will show, CNN and Fox News Channel news personnel maintained the character types created by the Bush administration and developed others on their own to sustain a simple narrative based on clearly opposing forces. Iraqi military officers also employed this tactic by calling Bush a "gangster" in a press conference, later explicitly comparing him to 1920s Chicago gangster Al Capone.

Former ABC foreign correspondent Malcolm Browne explains the appeal of character typing to news:

> Especially in America, we like to think of things in terms of good guys and bad guys. If one of the partners in a conflict is one that most people can identify with as a good guy, then you've got a situation in which it's possible to root for the home team. That's what a lot of news is about. We love to see everything in terms of black and white, right and wrong, truth versus lies. (quoted in Moeller 1999, 13–14)

CNN and Fox News Channel news personnel promoted the United States as the home team deserving of wholehearted support. One of the ways they achieved this was by describing the major participants in the conflict as recognizable character types that slid easily into the good guy/bad guy binary. For example, when referring to the initial reports of a grenade attack at Camp Pennsylvania, Aaron Brown speculated that it was tempting to report that a "bad guy" had breached the camp's security (CNN March 22). In another example, on March 24, when Rick Leventhal described a confrontation between U.S. troops and Iraqi soldiers armed with rocket-propelled grenades, Steve Doocy asked him how far he was from "the bad guys" (Fox News Channel March 24). Based on the war narrative established by the White House and the Department of Defense, CNN and Fox News Channel inserted a variety of figures into the roles of good guys and bad guys.

## The Good Guys

### Heroes

The list of the invasion's protagonists included President Bush and his so-called coalition of the willing, which was made up primarily of U.S. and British military personnel and token support from a number of smaller countries. Despite the fact that a number of nations were participating, the United States stood out in the coverage of both networks as the primary hero. The existence of the coalition was supposed to counter claims of U.S. unilateralism. Wolfowitz explicitly claimed on March 12 that "this is not going to be a unilateral action" (U.S. Department of Defense 2003b). Once the invasion started, Rumsfeld insisted that it was "not a unilateral action, as [it was] being characterized in the media" (U.S. Department of Defense 2003c). Openly unconvinced, Christiane Amanpour and retired colonel Mike Turner questioned the veracity of Rumsfeld's claim that the 2003 coalition was larger than that of 1991, and Judy Woodruff speculated that Bush was "anxious" to put out the message that the coalition was "broad" (CNN March 20).

Aside from these few instances of skepticism, most network journalists accepted official claims uncritically. Larry King repeated the assertion that the coalition was growing, and his guest, Senator John Kyle, asserted that "everyone" should assist in the reconstruction of Iraq (CNN March 20). On March 20, Fox News Channel reported that the coalition had apparently grown to forty-six nations, but the White House refused to release the comprehensive list of countries until March 25 (Fox News Channel March 20).[1] On March 21, Fox news personnel again reiterated that the coalition had grown, accepting the administration's numbers uncritically. One exchange between Neil Cavuto and David Christian repeated the claim with a degree of embellishment. Neil Cavuto commented that the coalition was continuing to grow, and David Christian replied that everyone wanted to "jump on board" (Fox News Channel March 21). Later, Christian painted the United States as the underdog by saying, "America is pulling something together that nobody thought they could do" (ibid.).

In spite of an initial focus on the coalition and its size, most commentary about the invasion's progress on the two networks addressed the efforts of the United States and not the British, whose troops constituted the second largest number of the war. Greta Van Susteren marveled at the "magnificent showing of American might," Mike Emanuel praised the "military might of the U.S.," and Tom McInerney reflected on "the power that we have" to go after terrorism and rogue states (Fox News Channel March 22). Geraldo Rivera remarked emotionally, "What the U.S. has embarked on is probably the most righteous mission that could possibly be conceived, and I just feel so strongly about that, I tell you, it gives me the chills" (ibid.). The coalition was the nominal protagonist, or good guy, but on Fox News Channel the United States clearly stood out as the military marvel responsible for the righteous deeds and presumably quick progress of the war.

---

1. In his March 20 press briefing, Ari Fleischer said the coalition was made up of thirty-five countries. He did not list all of the countries, except to say that the total population of the coalition countries was 1.18 billion, and the total gross domestic product of the countries was $21.7 trillion (White House 2003a). When the White House released the names of the countries on March 25, the number had grown to forty-six, the population was 1.23 billion, and the combined gross domestic product was $22 trillion (White House 2003d). Among the countries on the list were Afghanistan, Azerbaijan, El Salvador, Ethiopia, Micronesia, Palau, Rwanda, Tonga, and Uzbekistan.

In their descriptions of the U.S. president, Bush emerged as an active figure, a characterization that bolstered his image as a wartime president. The rhetoric on Fox News Channel typed him as a classic tough guy. Fox journalists described Bush as "gutsy," "aggressive," "independent," a "wartime leader," and "impressive" (Fox News Channel March 19–21). CNN news personnel typed Bush as an avenger by linking his actions in March 2003 to the events of September 11, 2001. For example, John King interpreted the "urgency" in Bush's address on March 19 as "very reminiscent to [*sic*] the morning after" September 11, 2001 (CNN March 19). At Fox News Channel, David Christian, Laurie Dhue, and David Asman also connected the Bush-led invasion with September 11 (Fox News Channel March 21 and 23). Their sentiments exposed their belief that an avenging Bush would create a safer United States.

The two networks typed the men and women who were actually doing the fighting somewhat differently. They described U.S. soldiers in the contexts of a quest for regime change, disarmament, and liberation. Not surprisingly, the revenge aspect of the war narrative also contributed to the typing of the soldiers. For example, as the 3–7 Cavalry rolled into Iraq, Walter Rodgers periodically interviewed the cavalry's commander, Captain Clay Lyle. Rodgers repeated Lyle's statement "It's payback time" from one of his pep talks to his troops (CNN March 20). However, instead of attributing rabidly vengeful characteristics to U.S. soldiers, CNN and Fox News Channel journalists preferred to type them as bona fide heroes with hearts as well as guns. That type of hero is the classic U.S. hero—tough yet selfless and aggressive only when pushed too far. Captain Lyle's comment incorrectly implied that terrorists had indeed pushed the United States too far and that U.S. soldiers were the heroes who had to push back.

Journalists at both networks constantly followed the script of the marketable concept (a vengeful yet humane U.S. military seeking to topple an evil regime) and the official narrative of the war; they never wavered in their vocal support for soldiers they constructed as altruistic heroes. Paula Zahn made her feelings on the subject clear when she said to Senator Johnson, "We salute your son's commitment" (CNN March 21). Embedded reporter Frank Buckley called the soldiers "professionals," while William Cohen, a *Moneyline* contributor, commented on his "awe and pride" (ibid.). As the NCAA basketball playoffs began, CNN's Aaron Brown commented that the soldiers were fighting so that college students could continue to play basketball (CNN March 22). Fox News Channel news personnel went a little farther in their typing of the U.S. military. Embedded reporter and former marine Oliver North stated that Fox News Channel portrayed soldiers "as they are—American heroes" (Fox News Channel

March 22). Fox News personnel referred to the soldiers as "brave," "impressive," courageous, "professionals," "spectacular fighters," "extremely brave," and "patriotic" (Fox News Channel March 19–23). Shepard Smith, E. D. Hill, and Steve Doocy thanked the soldiers for fighting and dying "for us and our safety" to keep "all of us free" (Fox News Channel March 20 and 24).

CNN and Fox News Channel personnel adopted the rhetoric of the Bush administration and transferred it to the soldiers, unquestioningly establishing a character type for them that was in keeping with the marketable concept and the quest for regime change, disarmament, and liberation.

### Sidekicks

Although soldiers were the heroes of the narrative at CNN and Fox News Channel, they had helpers. Both networks represented embedded reporters as acting in the service of the good guys to defeat the bad guys; they were the quintessential sidekicks. Sidekicks are helpers who do not get the glory, but they do get to go on the quest (and try to avoid being killed). As much as embedded reporters claimed allegiance to their news organizations, they functioned largely in the service of the military. They were helpers who followed along, emulating the "heroes" as much as they could. CNN and Fox News Channel journalists widely publicized the notion that embedded journalists played a vital role as go-betweens for the Pentagon and the troops on the ground. Douglas Kellner characterizes this differently, arguing that the reporters were nothing more than propagandists (2005, 64).

## The Bad Guys

### The Villain

CNN and Fox News Channel journalists fashioned villains to play the opposite numbers to the heroes they had constructed. Their repetitive claims that Saddam Hussein was a mentally unbalanced psychopath sustained and reinforced his image as a villain that had been in circulation since the early 1990s. For example, CNN's framed graphic of Hussein, which showed abbreviated data about his life and his conflicts, pointed out that the woman Hussein married in 1958 was his cousin, a fact that assaulted mainstream taboos about incest in the United States and painted Hussein as a regressive deviant. Additionally, because CNN used the framed graphic format most often to display and describe weapons, CNN's positioning of Hussein in the pantheon of weapons graphics connoted that he, too, was a weapon. The weapons of the bad guys posed the threat that made the marketable concept and war narrative possible, and the

Hussein-as-weapon graphic conflated the two Iraqi threats into one graphic representation of the villain of the war narrative.

## Henchmen

Like sidekicks, henchmen tend to be colorful characters. They perform at the pleasure of the villain, sometimes becoming victims of their employer. Recall the character of Bob, The Joker's loyal top henchman who winds up getting shot by his boss in Tim Burton's 1998 version of *Batman*. Unlike sidekicks, henchmen are one-dimensional fiends likely to be picked off as the movie progresses. Fox News journalists considered Mohammed Said al-Sahhaf, the Iraqi information minister, to be a henchman who was doing Hussein's bidding. According to CNN and Fox News Channel commentators, al-Sahhaf's role was to spread propaganda favorable to Hussein's cause. Upon hearing that the Iraqi Ministry of Information was using footage from embedded journalists as evidence of Iraqi military successes, Van Hipp called the information minister "a thug character out of a bad 'B' movie" (Fox News Channel March 23). For Van Hipp, al-Sahhaf was so completely a tool of Hussein that he had no agency at all; he was a caricature explained best by typing him as a character in a low-budget Hollywood film.

The other notable henchmen were practically invisible and always one-dimensional. One in particular came to the fore because of the threat he posed to U.S. troops. On March 21, William La Jeunesse, a Fox News Channel field correspondent in Kuwait, explained that Hussein had divided Iraq into five quadrants three weeks before the invasion. La Jeunesse pointed out that the southern quadrant—where many of the U.S. troops were concentrated—was under the control of Ali Hassan al-Majid, better known as "Chemical Ali" because of his history of using chemical weapons on the Kurds. La Jeunesse emphasized the gravity of the situation by saying, "If there's one person you don't want on the other side of the fence, it's Chemical Ali," a statement he repeated two more times that day (Fox News Channel March 21). Ken Pollack called Ali Hassan al-Majid "nothing but a thug" (CNN March 21). Ali Hassan al-Majid was one of twelve key members of Hussein's regime the Bush administration called the "Dirty Dozen" (Friedman 2003, 6). Considering that the film *The Dirty Dozen* encourages viewers to sympathize with twelve criminals, the nickname was a confusing way to describe Hussein's most dangerous men, but CNN and Fox News Channel journalists repeated it and reinforced the character typing of the Iraqis as henchmen.

Hussein's two sons, Uday and Qusay, also were members of the dozen, and CNN and Fox News Channel personnel occasionally singled them out as particularly fiendish henchmen. Simon Henderson recounted the deeds of Hussein's two "evil" sons, to which Aaron Brown replied sarcastically "Nice family" (CNN March 20). To journalists at both networks, Uday and Qusay Hussein were convenient and spiteful characters who proved that Saddam Hussein was genetically evil. Ken Pollack identified Qusay Hussein along with Ali Hassan al-Majid as the two most dangerous of Hussein's top men (CNN March 21). As CNN did with Hussein, Fox News Channel created framed graphics for Uday and Qusay. Uday Hussein's graphic linked him to an Iraqi newspaper and television network, and Qusay Hussein's graphic identified him as the "second most powerful [man] in Iraq" and the head of the "elite Republican Guards" (Fox News Channel March 21). Fox News Channel personnel accompanied the framed graphics with verbal accounts of the brothers' violent acts, highlighting the brothers-as-weapons connotation of the graphics.

The Iraqi military constituted the final group of henchmen whose alleged strength threatened the United States. However, network journalists needed to walk the fine line of relaying the threat of the Iraqi military while upholding the belief that the U.S. military was superior. Consequently, the conceptualization of the Iraqi military on CNN and Fox News Channel was an exercise in contradiction. On the one hand, personnel at both networks derided the Iraqi military frequently. On CNN, Iraqi soldiers and equipment were "1950s style," "old-fashioned," "not . . . good enough," not "well-equipped," lacking in training, "a nuisance," "a weak force," "not as good as [U.S.] soldiers," prone to surrender, "not very organized," in "disarray," and "notoriously inaccurate" (CNN March 20–24). Both CNN and Fox News Channel reporters used football analogies to describe the U.S. and Iraqi fighting forces: CNN's guest commentator Colonel David Hackworth compared the United States to the Dallas Cowboys and Iraq to a junior high team, and Fox's guest, Vietnam hero David Christian, compared Iraq to a high school team playing a professional team (CNN March 21; Fox News Channel March 21). Fox News described the Iraqi military as "not well-trained or equipped," "not very impressive," "limited," "not good," "just sitting there," prone to surrender, "not too significant," "not . . . organized," absent, structurally lacking, "broken," "falling apart," "not operating effectively," "desperate," "ineffectual," "ancient," "a ragtag bunch," pitiable, unshod, and "incapable" (Fox News Channel March 19–24). At the end of the day on March 20, Linda Vester declared, "The Iraqi army for the most part has given up" (Fox News Channel March 20). Later, Thomas McInerney said the war "always was a mismatch" (Fox News Channel March 21). If the plethora of demeaning descriptors was any indication of reality, the Iraqi military posed little threat to the United States.

On the other hand, CNN and Fox News Channel journalists also insisted that the Iraqis posed a substantial threat. CNN's Judy Woodruff, Barbara Starr, and Paula Zahn worried a great deal about the advanced equipment the Iraqis must have if they were prepared for the first round of bombings. Aaron Brown warned that the Iraqi forces might pull back and "make their stand" in Baghdad (CNN March 20). Walter Rodgers predicted that the Iraqis would "fight to the death." Miles O'Brien said that the Iraqi troops were motivated to fight because they were defending their homes (CNN March 22). At Fox News, William La Jeunesse frequently described Iraq's missiles and weapons, while Shepard Smith warned of land mines the Iraqis had laid (Fox News Channel March 20). Likewise, Bret Baier reported that the U.S. Air Force was concerned about the Iraqi military's capacity to use surface-to-air missiles. Major Garrett also expressed concern about "one place in Baghdad where Iraqi air defense systems are pretty good" (Fox News Channel March 21). Iraqi paramilitary units that engaged in hand-to-hand combat were highlighted in the coverage, as was the Ba'ath militia, which Greg Kelly described as a group of "very, very tenacious fighters" (ibid.). Add to that the concerns on CNN and Fox about Iraqi soldiers dressing in civilian clothes, using human shields, and simply "fighting dirty," in Shepard Smith's words, and the construction of the Iraqi military began to look quite muddled (Fox News Channel March 24).

For news personnel at both networks, the Iraqi Republican Guard was a more severe threat than the regular army. CNN journalists identified the "elite Republican Guard" as Hussein's "best paid" and "most trusted" force (CNN March 21). Ken Pollack said that the Republican Guard was 80,000 men strong, not counting the Special Republican Guard (CNN March 22). Retired general David Grange called the Medina Division of the Republican Guard the "first-class units" of the army (CNN March 24). Dennis Ross said that the Republican Guard had the "best equipment, money, [and] support" (Fox News Channel March 20). Furthermore, Fox's Rita Cosby said, the notorious chemical weapons red line around Baghdad was populated by a "circle of Republican Guard just waiting for us to come" (Fox News Channel March 21). Shepard Smith said that two divisions of the Republican Guard, in particular, constituted the potential "first spot of significant resistance" (Fox News Channel March 21). Robert Bevelacqua stated that the Republican Guard was "no match" for the United States, but he followed that by saying that a confrontation with its soldiers would be "by no means . . . a cakewalk." David Hunt deviated from the others, claiming that the ferocity of the Republican Guard was "urban legend" (Fox News Channel March 22). Additionally, the Special Republican Guard, essentially Hussein's force of private bodyguards, was described as a group of "glorified traffic cops"

(Fox News Channel March 21). News personnel at both networks collectively created a mixed characterization of this most feared Iraqi fighting force, as though admitting that the Iraqi military was strong would somehow diminish the might of the U.S. military.

This combination of bravado and fear resulted in a discursive balancing act. CNN and Fox News Channel journalists created a storyline that played up Iraqi military strength just enough to establish a threat that was credible enough to be part of the official war narrative and yet not diminish the superiority of the heroes. Remember that CNN and Fox News Channel news personnel had already written and publicized the conclusion of the war narrative. The recognition that Iraqis posed a threat not only validated the narrative, it also added drama without endangering the certainty of a coalition victory.

The villains of the narrative the networks constructed were by no means model citizens, and they easily fit into an understandable character type. Whether they were called goons, thugs, or just plain evil, CNN and Fox News Channel journalists positioned them within a pre-sold type that was both easily identifiable and easily communicated.

## False Heroes

CNN and Fox News Channel journalists situated France, Russia, U.S. antiwar protesters, and Arab television (particularly Al Jazeera) as bad guys, albeit a more deceitful sort of bad guy. CNN and especially Fox News Channel personnel felt betrayed by these groups. Yet because the members of these groups claimed that they were doing the right thing, they complicated the simple war narrative both networks clung to. As a way of handling this complication, network journalists constructed them as false heroes. The false hero, as constructed by Vladimir Propp, is the deceitful character who attempts to assume the role of the hero but who ultimately must be exposed and punished for the usurper that he or she is. False heroes are not true villains, but they are obstacles in the way of the hero's successful completion of the quest. Propp's term offers the most accurate way to describe how CNN and Fox News Channel positioned participants in the war narrative who claimed that Iraq posed no threat, characterized the United States as the aggressor and rogue nation instead of Iraq, and claimed to be on the side of righteousness. In short, everything the false heroes stood for threatened the marketable concept.

France's opposition to the U.S. invasion elicited the most frustration and vitriol from Fox News Channel. The network's journalists aggressively typed France as unreasonable and obstinate. They repeatedly wondered aloud how the French would respond if the invading troops found WMDs in Iraq. David Christian remarked, "The proof's in the pudding. We can tell [President Jacques] Chirac 'Here it is' and prove to the world 'Here are the sites. We told you so.'" Mark Ginsberg claimed that France would be "left in the dust" and "left with their pants down" if WMDs were found. He said that they would either "shamefully crawl their way back to our good graces or shamelessly oppose us to the bitter end." Tony Snow and Patti Ann Browne wondered how much proof the French needed to convince them of the good intentions of the United States and the existence of WMDs in Iraq. In response to speculation about where the coalition would ship recovered WMDs, Rita Cosby enthusiastically said: "France! What about France? Let's bring it all to France!" (Fox News Channel March 21). Fox News Channel journalists personalized France as a stubborn naysayer that would not listen to reason, and they compounded that portrayal by essentially accusing the French of conspiring with Hussein and his regime.

Fox journalists sought to discredit Chirac and all of France by associating them with Hussein, all the while ignoring the cordial ties of the United States to Hussein in the 1980s. The false hero must be exposed in the course of the narrative, and Tony Snow sought to do precisely that when he said, "The French have a lot to hide, that they have far more extensive commercial and even possibly strategic ties to Iraq" (Fox News Channel March 21). Senator John McCain echoed that claim two days later (Fox News Channel March 23). Ginsberg took this accusation further by stating that France had served as Hussein's "lawyer" for twenty years and that France's leaders had sided with Hussein and not the Iraqi people (Fox News Channel March 21).

When Chirac announced that the UN should be involved in the reconstruction of Iraq, Fox personnel sought to expose him as a weak-minded opportunist. David Christian likened France's leaders to "sheep . . . following the shepherd" into the reconstruction efforts. Brit Hume interpreted Chirac's move as France wanting "a piece of the action." Mort Kondracke called Chirac "arrogant," while Charles Krauthammer called him a saboteur eager to have veto power over a "war won with the blood and treasure of the Americans and British" (Fox News Channel March 21). David Asman said the French were "squirming to get back" into Iraq, and E. D. Hill exclaimed, "You know what? Stay home!" (Fox News Channel March 23 and 24).

Fox News Channel stood by this approach and applied it to another adversary when the U.S. State Department revealed that Russian companies had sold radar-jamming equipment and night-vision goggles to Iraq (Fournier 2003). The "slippery Russians," as Shepard Smith called them, joined the French as covert Hussein supporters (Fox News Channel March 23). Both Andrea Koppel of CNN and Julie Kirtz of Fox News Channel emphasized that the equipment made and supplied by the Russians was an immediate threat to U.S. troops (CNN March 23; Fox News Channel March 23). General McChrystal denied that claim at the March 24 Department of Defense briefing. Responding to a question about the sale of GPS jammers to Iraq, McChrystal responded, "In fact, we have been aware for some time of the possibility of GPS jammers being fielded. And what we've found is, through testing and through actual practice now, that they are not having a negative effect on the air campaign at this point" (U.S. Department of Defense 2003f). On the day of McChrystal's briefing, CNN's Wendell Goler added Iraqi civilians to the list of potential victims of the Russians, saying that the attempts of the United States to avoid hitting civilians with missiles and bombs could be impeded by the equipment Russia had supplied.

In spite of the fact that Russian President Vladimir Putin denied government involvement in the sale, John McCain implicated both Putin and the Russian government:

> I think it's disgraceful, I think that what this Russian government is doing in a number of areas [including] their brutal repression of the Chechnyan people, which to some degree we haven't given as much attention or concern about because of our desire to have good relations with the Russians. Let's not forget Mr. Putin's career was spent as a KGB agent, so his ideas of international dealings and . . . respect for human rights may be somewhat different from ours. (Fox News Channel March 23)

The evolution of the story at Fox News was striking. The culprits named for selling the equipment evolved from a global accusation against all "the Russians" to a more qualified subset, "Russian arms dealers," and finally, fourteen stories after the first report, to a much more accurate story about "one Russian company" (Fox News Channel March 23). Even after several Fox News Channel reporters attributed the arms sale to only one Russian company, network journalists peppered the coverage with statements that placed blame on Russia generally (Fox News Channel March 24). Even after the White House specifically stated that a Russian "company" or "companies" was responsible (White House 2003c), Alexander Haig reimplicated Putin late on March 24 with the statement, "If Putin wanted to do away with these shipments, they'd be done away with rather promptly" (Fox News Channel March 24). The aim of all of this was to discredit another leader who opposed Bush's plan to invade Iraq and who stood with Chirac in blocking the UN Resolution to authorize forceful disarmament (Xinhua General News Service 2003; Radyuhin 2003).

Fox News Channel journalists linked Putin to Chirac by raising questions about his ties to Hussein in an attempt to expose him as a traitor and thus an enemy of the United States. France and Russia—along with Germany, China, Cuba, Morocco, Cyprus, and others—were part of an international group that was opposed to the war (Agence France-Presse 2003). These countries presented themselves as conscientious objectors, but Fox News Channel personnel described them as deceitful opportunists who ultimately would harm the efforts of the United States to disarm Iraq and liberate the Iraqi people with minimal casualties.

Fox journalists did not confine this sort of character typing to international opposition to the war. Their treatment of the U.S. antiwar movement followed many of the same patterns. The typing of U.S. antiwar protesters is an example of striking difference between the two networks. Kellner's claim that "antiwar voices and protests were necessarily excluded in the profit-driven and pro-war atmosphere of media coverage" (2005, 68) is not quite accurate. In late March, antiwar protesters appeared multiple times on both CNN and Fox News Channel. Both networks reported on the U.S.-based antiwar protests in relation to seven primary topics: the significance of the rallies in a democracy, the size of the antiwar rallies, polls about public support for the war, the rationale for protesting, pro-war or pro-troop rallies or sentiment, protests that the networks characterized as "violent" or "disruptive," and the troops in Iraq. Each network's coverage used these topics to type the protesters, and Fox News Channel specifically used each to reject the premise that protesters were heroic in their exercise of democracy and freedom of speech.

According to Fox journalists, the protesters hated the United States and therefore did not practice democracy. The Fox reporters constructed a storyline that said that the number of protesters was minimal and that they represented the minority in public opinion about the war. They had no reasons for protesting other than to overthrow the U.S. government. They were violent and encouraged violence, which contradicted their verbal appeals for peace. They said they wanted democracy but they supported Hussein—not the U.S. troops—and did not believe in bringing democracy to the oppressed Iraqis. They claimed to speak truth to power heroically, but they merely represented another evil power. According to Fox journalists, the U.S. citizens who exercised their democratic right to oppose a war they considered unjust were no less than traitors.

CNN journalists also disagreed with the actions of the protesters, but they handled their coverage of antiwar events much differently than the Fox network did. Instead of vilifying the protesters, they co-opted their message by emphasizing how antiwar activity demonstrated the tolerance of U.S. democracy. Aaron Brown asserted that he and his colleagues at CNN were "great believers in the right to demonstrate," and both he and Judy Woodruff gave small speeches on the superiority of U.S.-style democracy, of which protests were a vital part (CNN March 20–22). Guest commentator Senator Kay Bailey Hutchison stated that the right to protest was "in line with freedom of speech" (CNN March 21). Fiske notes how statements made in support of protesters illustrate the process of "inoculation" in news programming, an incorporation of radical voices into an official narrative so that the opposition actually fortifies the status quo. In this process, journalists first accord the oppositional speech minimal importance so that the rhetoric can subvert the dominant ideology without exacting any real damage. In addition, the representatives of the news media "[speak] the final 'truth,'" an act that frames oppositional speech from a specific viewpoint (Fiske 1987, 290–291). CNN journalists allowed protesters to have their sound bites and then defused the power of antiwar speech by applauding a tolerant democracy.

CNN's practice of "ideological containment" was just as problematic but less combative than the tactics presented on Fox News Channel. The Fox network did not feature the protesters as much as CNN. By way of explanation, Shepard Smith announced that Fox News Channel was keeping coverage of the antiwar protests "limited" in order to keep the network's overall coverage "fair and balanced" (Fox News Channel March 20). However, when the network did focus on the protesters, its journalists failed to uphold the right of U.S. citizens to protest in a time of war. Although guest commentator Mayor Rudolph Giuliani called the protests "a necessary part of democracy," an exchange between Fox's John Kasich and Congressman Greg Meeks became heated when the subject of democracy arose. In response to Kasich's "stunned and mystified" reaction to the protests, Congressman Meeks expressed his hopes for a democracy in which people had the right to dissent. Kasich responded by telling the protesters to "shut up," prompting Congressman Meeks to accuse him of not believing in democracy (Fox News Channel March 22). Alan Colmes, the self-identified "liberal" of the program *Hannity & Colmes,* voiced the most prominent defense of free speech on Fox News Channel with this remark: "People have the right to [protest], and certainly that's not in debate, but I think some of these people feel they have to prove they have the right, but they don't have the obligation" (Fox News Channel March 24). If Fox journalists gave any attention to the antiwar protests at all, it was only to imply or even to insist that they should not be speaking. The network gave no time to the real news of the antiwar protest stories, which would necessarily have focused on the issues the protestors were raising and the points they were making.

Reports of the size of the protests also produced differing coverage at CNN and Fox News. CNN's Aaron Brown characterized the size of rallies in Chicago and San Francisco as "significant," though Wolf Blitzer said the Chicago rally was smaller than the antiwar protests at the 1968 Democratic National Convention in Chicago (CNN March 20). Maria Hinojosa estimated a New York City protest to be twenty blocks long and said that approximately 200,000 people were there (CNN March 22). In contrast, Shepard Smith asserted that the antiwar protests were "not enormously large," and Brian Kilmeade called all of the protests in the United States "small" (Fox News Channel March 20 and 21). The pro-troop/pro-war rallies covered by CNN and Fox News Channel were substantially smaller than the antiwar protests, but both networks covered them in the self-proclaimed pursuit of objectivity.

CNN and Fox News Channel both discussed opinion polls, which provided some degree of context in their coverage of the war, but Fox News Channel used polls to dismiss the validity of dissenting views. Fox personalities repeatedly noted the CNN/*U.S.A Today*/Gallup polls that showed that 72 percent of people in the United States supported the war, and they often cited the polls during or after stories about antiwar protesters (Newport 2003). A statement that Rebecca Gomez made typified the tone of Fox News Channel's coverage. Gomez covered the protests (as did the Latinas on CNN—Maria Hinojosa and Teresa Gutierrez), and she introduced one of her pieces with the statement that "the antiwar crowd refuses to acknowledge the polls" (Fox News Channel March 22). In her perception, polls should dictate opinions rather than reflect them. This (mis)perception was characteristic of the attitude of Fox News Channel's personnel: the protests were irrelevant annoyances because the polls said so.

However, Fox News journalists quickly decided that the protests were indeed newsworthy as reports of violence and disruption became part of the story. The issue of disruptive protests led the network's personnel to label the demonstrators a "safety hazard," further discrediting their intentions. Journalists consistently declared that the protests were a danger. Rebecca Gomez claimed that protesters "nearly attacked" her and her crew; she interpreted this as evidence that the protesters "seemed to want to take their anger out on someone" (Fox News Channel March 22). Reporting on a protest in Washington, D.C., Shepard Smith said the demonstrators were "making a mess of the morning commute" and were keeping firefighters "from answering emergency calls." He also declared San Francisco to be in a state of "absolute anarchy" (Fox News Channel March 20). Brian Kilmeade said protesters were "out of control" and were "breaking the law" (Fox News Channel March 21). Bill Cowan called Market Street "ground zero" in the protesters' "mission to paralyze downtown"

Chicago and frustrate "innocent drivers" (ibid.). Bob Sellers highlighted arrests in San Francisco and police chasing protestors at a New York City rally (Fox News Channel March 22). Rebecca Gomez called the crowd at the rally she covered in New York "out of order" and raised the concern that marchers "diverted limited resources from stopping possible terrorist strikes" in a "city already hurt financially by 9/11" (Fox News Channel March 22). John Kasich, Miami police commissioner John Timoney, Brian Kilmeade, and Linda Vester all argued that protesters diverted resources from homeland security (Fox News Channel March 22–24). According to Fox News Channel, the protesters were violent and posed dire threats to the safety of the United States—a claim that linked the protesters to terrorists.

CNN's reporting of the protests was much more balanced. It reported that 1,000 arrests and various run-ins with police had occurred at protests in New York City and San Francisco, but both Aaron Brown and Wolf Blitzer pronounced the protests "peaceful" (CNN March 20 and 21). Blitzer even prefaced a story about the protesters in New York City with the line "Freedom of speech led to urban gridlock," emphasizing the disturbance but linking it to a constitutional right (CNN March 22).

Fox News Channel journalists insisted that the "disloyalty" of protesters was particularly disturbing a time when they perceived unity to be the true sign of patriotism. Personalities on Fox News Channel achieved this by characterizing protesters as opposed to the troops, freedom, and democracy and supportive of terrorists. The issue of supporting the troops was less black and white on CNN. Aaron Brown stated that one result of the Vietnam War was that "we don't blame soldiers" for policy decisions (CNN March 21). With that, Brown defended the protesters by distinguishing anti-troop sentiment from anti-policy sentiment. Fox News Channel journalists were unable to tolerate any distinction between criticism of government policy and criticism of the troops. Guest commentator Scott O'Grady, a former Air Force Captain who had been shot down over Bosnia and later rescued, suggested that the protesters realize "that there are evil people in this world, and this is a just war, and we need to be supporting" the troops (Fox News Channel March 23). Guest Jeffrey Zaun, a former POW of the 1991 Persian Gulf War, took this a bit further, claiming that the protesters were "insulting" the troops (Fox News Channel March 23). Rebecca Gomez summed up the tone of Fox News Channel by saying, "Some of the protesters claim they do support the troops, they just don't support the war, and they seem to think that they can do both" (Fox News Channel March 22). Her statement reflected the kind of consciously uncomplicated logic that Fox News Channel typically used when speaking of the protesters.

Accusations that the protesters were disloyal grew increasingly severe on Fox News Channel over the course of the first five days of the invasion. By claiming that the protesters stood "against any war for the liberation of Iraq," Bob Sellers simultaneously supported the Bush administration's official reasons for the invasion and typed the protesters as obstacles in the quest for freedom (Fox News Channel March 22). Shepard Smith hinted at an insidious conspiracy when he declared that the antiwar protests were "part of a synchronized movement to stage protests" (Fox News Channel March 20). His statement was in keeping with the network's numerous attempts to link antiwar protesters to organizations that Fox journalists deemed to be subversive—a tactic similar to the network's typing of France and Russia as disloyal to the United States because they criticized the war. Tony Snow modified the commonly used phrase "pockets of resistance," a way to describe what U.S. troops were up against in Iraq, to "little pockets of protest," his way of at once describing and minimizing the impact of antiwar activities in the United States and the United Kingdom (Fox News Channel March 22). David Asman argued that the protesters were vulnerable to Iraqi propaganda, and one guest claimed that the protesters' rhetoric "play[ed] into the hands of our enemies" (Fox News Channel March 22 and 24). In addition to calling "most" of the protesters "stupid," Fred Barnes accused them of being "objectively . . . pro-Saddam"; in disgust, Tony Snow replied, "Enough of them" (Fox News Channel March 22).

E. D. Hill faulted other news outlets for not disclosing the financial backers of protests. "What you find out when you start looking at who's financing them," she revealed, "is that many, many of the organizations are linked to Castro, to, to dictators" (Fox News Channel March 21). Former Republican senator and Hollywood actor Fred Thompson also accused the protesters of acting in the service of terrorists. He claimed,

> This whole antiwar protest so-called movement is being used by international groups and so forth that are virulently anti-American, anti-west, anti–free enterprise, who use this in order to move from a demonstration . . . to stopping things and attacking really some of the same sites that the terrorists would. (Fox News Channel March 24)

Thompson was also concerned about "organizations that are international in scope," a statement that fed into the overall xenophobic tone of Fox News Channel news coverage. David Horowitz, president of the Center for the Study of Popular Culture, combined all of the previous claims into one large accusation. He denied the existence of a peace movement, calling it instead "hardcore . . . Communist." According to Horowitz, the protesters, who "hate America" and spewed "Marxist absurdities," were running a "sabotage campaign" that endorsed "commit[ing] violence" and "smash[ing] the state." He wrapped up by accusing the protesters of being involved with "Palestinian terrorist organizations and terrorist support organizations" as well as with North Korea (Fox News Channel March 24).

In short, in Fox News Channel's estimation, the protesters were not to be trusted. Fox journalists and guests typed the protesters as false heroes with even more fervor than they exhibited when typing France and Russia. In contrast, CNN reporters and guests did not resort to the type of inflammatory rhetoric that was a mainstay on Fox News Channel; CNN's approach was to contain the message of the protestors by pointing to the right of citizens in a democracy to free speech rather than focusing on the issues the protestors were raising.

CNN altered this approach somewhat in its reporting about Arab television. Both networks, in fact, joined forces in the common project of constructing another false hero: the Al Jazeera satellite news network. CNN and Fox News Channel news personnel presented a generally negative view of Arab television from the outset, accusing the various Arab networks of being dangerous, anti-American, and propagandistic. CNN's Wolf Blitzer discussed the possibility that "inflammatory" images on Arab television "could generate more terrorism" (CNN March 22). Fox's William La Jeunesse dwelled on the theme of inflamed emotions when he claimed that Arab media was "playing this war" to incite emotional responses (Fox News Channel March 24). The general sense that Arab television networks had aligned against the United States was prevalent on Fox News Channel. Linda Vester reported that Arab networks were pleased that the Iraqi military could "poke the American military in the eye" by using their embedded footage against them (Fox News Channel March 23). Simon Marks relayed the message of Jordan's media that the war was "an unjust war of aggression . . . designed to enslave the Iraqi people" and "seize Iraq's oil" (Fox News Channel March 24). The overwhelming description of Arab television on CNN and Fox News Channel, then, was that it was a vehicle for pro-Hussein propaganda.

The propaganda value of images and stories on Arab television was not lost on CNN and Fox News Channel news presenters. In the context of a discussion of the slow rate of information disclosure at CENTCOM, Jane Perlez said that some information was available from Arab television, but she explained, "It's not information. It is an Arab version of how this war is going" (CNN March 24). Whenever Fox journalists mentioned Arab television, they quickly noted that the information it conveyed could be false. For example, when Rita Cosby reported news of Iraqi resistance and U.S. deaths, she immediately said, "But of course that's all coming from Arab television and the Iraqi information minister. U.S. officials are saying no such thing" (Fox News Channel March 22). Steve Harrigan emphasized that there were "two different wars, two very different versions of how this war is going depending on which channel you're watching"; later, he explicitly stated that Arab television was televising a "radically different war" (Fox News Channel March 24). David Christian called Al Jazeera and the Abu Dhabi television network "propagandists" (Fox News Channel March 23).

Because CNN no longer had a stranglehold on 24-hour satellite news as it did in 1991, both CNN and Fox had to contend with competing images of war in 2003. The greatest foe of both networks in 2003 was not the whole of Arab television but rather the network known as "the Arab world's CNN"—Al Jazeera (El-Nawawy and Iskandar 2002, 24).[2] Al Jazeera, the only independent network in the Middle East, was founded in 1996. The network's Web site claims that it brings a balance to news reporting by giving voice to new perspectives, particularly those from the developing world, in a way that challenges the current flow of news based on a developed-world point of view. Al Jazeera feels that it offers a more balanced perspective that has "changed the face of the news" through "accurate, impartial and objective reporting" (Al Jazeera 2008). According to El-Nawawy and Iskandar, the satellite network's philosophy is "built

---

2. When the BBC's Arabic television division lost its Saudi funding in 1996, most of the staff was hired by executives who were creating a new satellite news network, Al Jazeera. Al Jazeera, in turn, received $140 million in funding from the emir of Qatar, Sheikh Hamad bin Khalifa Al-Thani. Sheikh Hamad conceived of Al Jazeera as an independent and nonpartisan news organization operating without government supervision (El-Nawawy and Iskandar 2002, 31, 33). Al Jazeera, the "first truly liberal, independent satellite channel in Middle Eastern history," has between 35 and 70 million Arab viewers, depending on the source (Bhatnagar 2003; DemocracyNow.org 2005; Fahmy and Johnson 2007; NOW on PBS 2005; Sharkey 2004; Wu 1999). The network made its first major impression on the U.S. media with its exclusives from Afghanistan after September 11, 2001. Soon after that, Al Jazeera entered into contractual video-sharing relationships with CNN, ABC News, the BBC, and Germany's ZDF (Zednik 2002). Al Jazeera dissolved its partnership with CNN in January 2002 after CNN obtained and aired a videotape of Osama bin Laden that Al Jazeera had chosen not to air (El-Nawawy and Iskandar 2002, 169). Nevertheless, in March 2003, every U.S. news network used footage from Al Jazeera, whose cameras in Baghdad were an inadvertent "intelligence tool" for the United States, making it possible for U.S. officials to assess the damage U.S. bombs had done (Sharkey 2003).

on demonstrating how objectivity can be attained only if all subjective views and opinions on any issue are presented and aired" (2002, 27). The fact that Al Jazeera's news programming has generated criticism from both the United States and from Middle Eastern governments lends credence to the notion that its news coverage is balanced.

Despite Al Jazeera's philosophy of balanced news coverage, CNN and Fox News Channel campaigned against the independent network as though it were a state-run institution like Iraqi television. While Al Jazeera prided itself on being the sole example of free media in the Middle East and was intent on presenting the 2003 invasion as it really happened, CNN and Fox News Channel sought to expose it as a mouthpiece for anti-U.S. rhetoric and propaganda. As with their coverage of France and Russia when they questioned U.S. policy and their coverage of U.S. citizens who protested the war, CNN and Fox News Channel on-air personalities positioned Al Jazeera as a false hero—a sensationalist liar and propagandist with ties to the Hussein regime.

Surprisingly, Fox News Channel was not without its praise for Al Jazeera. John Gibson referred to it as "one of the more respectable television networks in the Arab world" (Fox News Channel March 24). However, when Carl Bernstein called Al Jazeera "a great thing" for the Arab world, David Asman let out a resounding "Whoa!" Asman admitted that Al Jazeera did "open their airways to [the U.S.] point of view," but Bernstein's claim that Al Jazeera broke "the propaganda machine" of several Middle Eastern countries was not the prevailing view among Fox News Channel's staff (Fox News Channel March 23). William La Jeunesse described Al Jazeera's war coverage as "the most anti-war, pro-Iraq [coverage] that we have seen"; he went on to accuse the network of "basically reporting what the government is saying there with very little skepticism" and of "minimizing the explanations the U.S. has had for why this war is taking place" (Fox News Channel March 22). Discussing Al Jazeera's tour of Baghdad hospitals to assess the status of civilian casualties, Fox's Mort Kondracke stated that the network would "not interpret [the visit] in a way that is going to be favorable to the U.S." (Fox News Channel March 21). David Asman announced that Al Jazeera was not showing the Safwan "liberation" video (see chapter 2) because it did not want to show footage of "the people who are happy to be liberated in Iraq." Asman reinforced Tony Snow's assertion that Al Jazeera's coverage had been "very light on reporting of allied victories and rather heavy on Iraqi propaganda" (Fox News Channel March 23).

Allegations of propaganda increased as the war intensified four days into the invasion. On March 23, Al Jazeera raised the ire of the U.S. Department of Defense, CNN, and Fox News Channel by airing a six-minute tape of U.S.

POWs provided by Iraq TV. An Al Jazeera announcer prefaced the network's first broadcast of the video with the following statement: "We have to share this evidence with you because the Iraqis are giving it out as evidence of what they say—that they have killed U.S. soldiers" (Carter 2003). Al Jazeera aired the tape until the Pentagon requested that it stop so the military could notify the families of the soldiers. The reaction to the video was forceful and swift. One day after it aired, the New York Stock Exchange banned Al Jazeera financial analysts from the trading floor (Kolodzy, Ricks, and Rosen 2003). CNN and Fox News Channel journalists voiced their equally negative reaction to the video frequently and descriptively. On the day the POW video aired, Major Garrett accused Al Jazeera of maintaining "very close coordination" with state-run Iraqi television (Fox News Channel March 23).

In the view of CNN and Fox News Channel, Al Jazeera had committed a foul offense, and CNN chose to confront the Arab network. The day after Al Jazeera aired the POW tape, CNN's Aaron Brown conducted an interview with Al Jazeera's chief Washington correspondent, Hafez Al-Miraz. In the interview, a visibly agitated Brown departed from CNN's more measured style and adopted a stance more typical of a belligerent Fox news anchor. Al-Miraz said that he felt that it was unfortunate that Al Jazeera had initially rushed the footage to the airwaves without editing it and said that the network had quickly realized that it needed to be edited and had done so to protect viewers from the most gruesome parts of the footage. He added that Sky News and European networks had shown the video unedited, implying that some Western-based networks had not shown the same caution that Al Jazeera had after an initial misstep. He also stated that Al Jazeera had complied with the Pentagon's request to stop airing the tape until the families of the soldiers could be notified. But he also noted that half an hour after Al Jazeera stopped using the footage, a CNN reporter had released the names of three POWs in the footage and had been reprimanded by the Pentagon. In a series of tense exchanges that followed, Brown grew more indignant, interrupted Al-Miraz repeatedly, and dominated the interview, hardly giving Al-Miraz an opportunity to speak. Brown asked Al-Miraz if the reporter who had revealed the identities of the POWs before families could be notified was on CNN International or the CNN domestic network. When Al-Miraz replied that the incident occurred on CNN domestic, Brown said, "Excuse me, because as you know . . . that's a very different audience and a very different issue there." But he would not accept that same argument when Al-Miraz insisted that as an "Arabic language news network," the POW footage was for Al Jazeera's "own audience" (CNN March 23). Though at the end of the interview Brown made the exact same argument that Al-Miraz had made at the beginning of the interview—that the intended audience was the key in determining the propriety of images—his line of questioning fixated on the broadcasting of the tape without addressing the issues that Al-Miraz had raised. Had Brown engaged with Al-Miraz on any of the pertinent issues—the representation of war, the wide circulation of the tape beyond the Middle East, the repeated images of Iraqi POWs on U.S. television, and CNN's own history of televising dead U.S. soldiers—the conversation may have yielded a more nuanced view of the role of Al Jazeera. Instead, Brown constructed Al Jazeera as an irresponsible tabloid pandering to morbid curiosity.

While CNN journalists focused on the graphic detail of the POW video and what they believed to be the poor judgment Al Jazeera exercised in airing it, Fox News Channel journalists associated the entire situation with propagandistic aims. Disregarding Fox News Channel's (and CNN's) repeated use of images of Iraqi POWs, Brian Kilmeade asked, "What about Al Jazeera? Why would you air something over and over again, every half-hour, every forty-five minutes? . . . How does that help any news organization to air captive people?" Steve Doocy replied, "I'll tell you why Al Jazeera's running it. 'Cause they're not on our side." As if to affirm that view, Chris Jumpelt asserted that the POW video was a "morale booster" for Iraqis who viewed the liberation as an invasion (Fox News Channel March 24). For Fox News Channel personnel, Al Jazeera's actions represented more than sensationalism; the POW video incident "exposed" Al Jazeera as a network intent on helping state-run Iraq television spread propagandistic and violent images.

The treatment of Al Jazeera at CNN and Fox News revealed as much about the two networks as it did about Al Jazeera. In other words, Al Jazeera's position within the war narrative hinged on how each network defined its own place within the profession of journalism. Al Jazeera posed a threat to both networks in two major ways. First, Al Jazeera threatened the power over distribution that CNN and Fox News Channel enjoyed. Fiske writes that the immediacy that defines television news privileges the "large multinational news corporations" that have the means to distribute news quickly via satellite. This control over distribution means that internationally distributed news has been, as Fiske puts it, "white," meaning that most network news delivers only the perspective of the white-dominated Western world (Fiske 1987, 289). Thus, Al Jazeera's satellite distribution challenged the international reach and dominance of Western megacorporation Time Warner—via CNN International—and News Corporation—via Sky News. That challenge increased considerably with the debut of Al Jazeera's English-language channel.

Second, CNN and Fox News Channel represented a considerably different approach to the war than Al Jazeera did, and the POW tape brought that difference to a head. Al Jazeera's conception of its audience and of the role of a free press during wartime simply did not translate to the language of commercial U.S. media or to the logic of high concept. In their statements, CNN and Fox News Channel news personnel restricted their understanding of Al Jazeera to visceral reactions and nationalistic fervor. Their concern centered on public relations and not on the material reality of a large-scale invasion. In limiting the dialogue about the POW tape, CNN and Fox News Channel personnel disregarded Al Jazeera's context. They rejected the stated goals of Al Jazeera— to shed the appearance of "war-as-videogames" and report as "independent news media," as Al-Miraz put it—and emphasized their disgust and feelings of betrayal at Al Jazeera's refusal to behave as commercial U.S. news networks would (CNN March 23). Though both networks continued to use images from Al Jazeera's cameras, they assigned a character type to the Arab network that would not be redeemed.

## Conclusion

CNN and Fox News journalists and their on-air guests constructed heroes (false or otherwise), villains, henchmen, and sidekicks to perform specific roles in the narrativized invasion of Iraq. CNN and Fox News Channel used specific participants to fill character types that were easy to communicate and easy to appreciate—two fundamentals of high concept. How viewers negotiate and understand a character transparently labeled "villain" has much to do with their experiences with other villain types and with how these villains have typically been portrayed in the news. This type of identification ranges from images of particular people—Hussein, for example—to entire groups of people linked by culture and religion. Like the typed characters in high-concept films, the characters CNN and Fox News personnel created represented sensibilities and ways of being that were simplistic and lacked nuance. There was little or no true analysis of the news on the two networks during the first five days of the war, but there was much talk about how various elements of news stories fit the character types that served the official U.S. government's narrative of the war.

The simplicity of the characters helped shape a more exciting narrative that would attract bigger audiences. For CNN and Fox News Channel, in-house stars (embedded journalists) personalized the networks' role in the war coverage and helped ratings, while established stars like Bush and Hussein simplified the networks' explanation of and justification for the war. The character types the networks implemented also provided the kind of comfort available in formulas. Comforting formulas attempt to restrict meaning, and in news programming, that translates to simplification and decontextualization—strategies that do a disservice to citizens who wish to be informed in as comprehensive a manner as possible. Pre-sold genres, stars, and character types made the war narrative more compelling to watch, and they reinforced and in some cases intensified the official narrative emanating from the Bush administration.

# 5

## *The Look and Sound of*
## *High-Concept War Coverage*

During a live firefight at Umm Qasr on March 23, former Deputy Assistant
Secretary of the Army Van Hipp and military analyst David Christian commented on the events for Fox News Channel. Eager for a swift and decisive
end to the firefight that seemed to drag on, Van Hipp remarked, "Hopefully
we're all going to get to see the effects of a JDAM [Joint Direct Attack Muni-
tion] bomb on that location right now." David Christian added, "Hopefully
the British cameraman has pointed the camera in the right direction so that
we do have the pictures of the actual dropping of the bombs. Because it would
be somewhat sad to have viewed this for two hours, this sight, and he's got the
camera pointing in the wrong direction" (Fox News Channel March 23). In
one sense, the camera linked the commentators to a version of reality, but that
reality required a payoff—a climax that, regardless of consequences, needed to
be seen on television.

As stylized as television news may be, the stories it airs still purport to
be grounded in objective reality. Ellen Seiter writes that the camera has been
linked so convincingly to the idea of objectivity that viewers simply do not
perceive that the images are constructed (1992, 38). For this reason, a semiotic approach to the look and sound of high-concept news is fitting. Seiter
argues,

> All communication is partial, motivated, conventional, and "biased," even
> those forms such as print journalism that are founded on a reputation for

truth-seeking and attempt to convey the impression of reliability. The study of semiotics insists that we should discern the distinctive ways of producing and combining signs practiced by particular kinds of television, in particular places, and at a particular point in time, because these codes are inseparable from the "reality" of media communication. (39)

While most research on the visuals of television news focuses on viewer cognition and recall of information, semiotics contends strictly with the production of meaning through signs. The approach involves only one facet of the media artifact, but when placed within the context of high concept and all that it entails, it allows us to expand our understanding of the war story that CNN and Fox News Channel created and transmitted visually and sonically. It also affords us the opportunity to understand how "violence, vulnerability, fear and uncertainty" mesh with technology to alter "the character and function of representation" (Giroux 2006a, 20).

Visuals are privileged elements of the high-concept style of filmmaking, and aural elements other than music are not a significant facet of Wyatt's conceptualization of high concept. This limitation hampers our ability to arrive at a holistic understanding of high concept in an audiovisual medium. An investigation of sound is perfectly in line with the stylistic preoccupations of high concept, however. Whether it takes the form of music, comes from the video footage itself, or manifests artificially as sound effects, sound figures prominently and strategically into CNN's and Fox News Channel's war coverage.

One sonic element—music—is an essential facet of high-concept films because it operates in the service of the marketable concept. Music comments on the characters' lifestyles in high-concept films more than it comments on the narrative, and musical interludes often seem excessive. In other words, music at times overwhelms the narrative. When music and other sounds are excessive, they call such a high degree of attention to themselves that they detract from (and in some cases halt) the narrative. Wyatt attributes these moments of excess not to the personal vision of an auteur but to the marketplace itself; he writes "the logic of the marketplace is clearly the author of the style" (Wyatt 1994). Nonmusical sounds (both diegetic—emanating from within the world of the story—and extradiegetic) also support the marketable concept and can be excessive. These principles also apply to television, where sound is manipulated for commercial gain.

In his influential article "Television/Sound," Rick Altman argues that television soundtracks in the United States are designed to recall the attention of distracted viewers and keep them engaged (1986, 39).[1] In negotiating the tension between Raymond Williams's (1974) programming "flow" and Altman's "household flow," the soundtrack serves the goals of the television industry and accommodates the daily routine of the television audience (Altman 1986, 40). With this in mind, an analysis of high concept in television news programs should consider more than just the visual look of the product; it should consider how news programs use sound as part of the "hook" to keep viewers engaged.

The sound and the look of television news programming need to be analyzed in the context of their role in the overall coverage, however. In her examination of coverage of the 1991 Persian Gulf War, for instance, Melani McAlister cautions against emphasizing form for form's sake:

> We must, of course, ask what role the hyperextension of visual images played, but in order to understand that, we need to know the intertexts and contexts of Gulf War television—how it engaged with other narratives about history and identity to envision the world it was helping to create. (2005, 245)

If we divorce the audiovisual components from the entirety of the coverage, then we risk decontextualizing our analysis of network news coverage in the same way that the networks decontextualize the news.

This chapter will analyze the visual and sonic events from coverage of the 2003 invasion of Iraq on CNN and Fox News Channel, including title sequences, the bombing campaigns, news graphics, the machinery of war, and images provided by embedded journalists. The complex interaction of select images and sounds from the coverage illuminates the hyperstylization of television news as well as the dedication of network presentation style to the marketable concept.

### Understanding Images and Sounds

Ellen Seiter writes that one major component of television's importance stems from its use of all five of Christian Metz's channels of communication at once: image, written language, voice, music, and sound effects (Seiter 1992, 45; Metz

---

1. Not everyone agrees that all television viewers are distracted. For example, John Caldwell (2000) argues that "videophiles"—"informed and motivated viewers"—constitute a large part of the viewing audience.

1999). In order to apply an analysis of these channels to television, Seiter uses the term "graphics" instead of "written language" (1992, 43). Significantly, the five channels of communication highlight the central role of sound on television. As Wyatt's term "the look" indicates, considerations of the visual realm often and unfortunately eclipse the auditory realm in analysis of television as a medium of communication. Metz writes that sound is considered secondary to the "visible object" precisely because it is regarded as a "nonobject" (1999, 358). Film theorists' discussions of sound define it in relation to visuals, as in the case of the term "off-screen sound" (357); they see sound as a substantial but lesser presence than images. In this limited conception, sound lacks "authenticity" because it is "incomplete" without the image (Belton 1999, 378). Semiotic analysis underscores the signifying powers of both visuals and sounds.

News programming creates a strong connection to reality by positioning images as indexical signs, or signs that stress a "material connection between signifier and signified" (Seiter 1992, 36). Seiter points out that when television journalist Andrea Mitchell appeared to be standing in front of the White House but actually was not, the image was a complex fabrication instead of an indexical sign (37). Semiotic analysis reveals that news is not "pure information" even if it does consist of indexical signs; rather, the news, like anything else on television, consists of the "active production of signs" based on social and cultural conventions. Seiter cautions us to treat television as a collection of signs and not as referents because the referents of what we see and hear on television are "impossible to verify" (38).

John Fiske argues that in order to achieve realism, television must be faithful to the "physical, sensually perceived details of the external world." Like Seiter, he notes the importance of indexical signs. He contends that television's appeal to realism must also involve an allegiance to the "values of the dominant ideology" and asserts that the "constructedness" of the signifiers of realism on television inextricably bind dominant ideology to the so-called "objective world of 'reality'" (Fiske 1987, 36).

The relationship between a dominant ideology and an objectively observed reality on television is illustrated by Brian McNair's claim that television news is the "most 'trusted'" source of news because of the "denotative accuracy" of photography and videography (1998, 67). The belief that recorded images simply transmit reality is a cornerstone of the claim that television news is objective. This problematic notion extends to beliefs about the sound one hears on television as well. Steve Wurtzler writes of television's need to match the source of sounds with "image events." For instance, when images of a correspondent or a location are not available—as they were not in some reports from embedded journalists—the sounds of voices or events are aired alongside still photographs or graphics. Wurtzler argues that these types of images serve as "conventionalized representations of a sound source" (1992, 92). No moving images are present to authenticate the sound we hear, so the stills compensate for that absence. For John Belton, the resulting function of the soundtrack, "duplicate[s] . . . the sound of an image," not the sound "of the world" (Belton 1992, 379). Belton assesses sound in the field of cinema, so to translate this view of sonic reality to television news is slightly problematic. If one subscribes to the idea that the news can represent reality objectively (and news networks support that idea wholeheartedly), then the sounds emanating from the footage should be just as "objective" as the images are. Tampering with "live" sounds compromises journalistic objectivity, and the addition of sound effects and music necessarily raises similar concerns.

Sound operates under very particular circumstances on television. Rick Altman outlines four primary functions of the television soundtrack. First, "auditor[s]" should have no doubt that the soundtrack will keep them informed while they are within earshot of the television but not looking at the screen. Second, auditors should be assured that the soundtrack will alert them when something significant transpires. Musical cues and sound effects work to this end on television news. Third, there should be no surprises in sound or subject matter throughout the programming day. (Here, Altman essentially describes formula and format.) The final function of the soundtrack is to "provide desired information, events, or emotions" throughout the day (Altman 1986, 42–43).

John Ellis differentiates television sound from sound on film—a medium whose viewing circumstances are more restrictive than the household setting. The circumstances of watching television in a domestic space are markedly different from the setting when one watches films in a movie theater; a television viewer at home can continue to "watch" television without being in front of it. Ellis places sound on television on a higher level than the images. Like Altman, Ellis notes that in the "regime of viewing television," sound "holds attention more consistently than the image." Ellis believes that television news relies on sound "as the major carrier of information" (Ellis 1999, 385–387). Most of the sound on television news comes in the form of verbal communication. Sound effects, silence, and music cannot compete with the dominance of the spoken word (or the images and graphics) in the genre of television news. Nevertheless, nonverbal sounds contribute to style, narrative, and marketability in television news.

Music is one major sonic element that assumes special significance in television programs that attempt to convey realism. John Corner notes that the two primary functions of music on television are to provide thematic support and organizational support. Music can signal time, place, and mood as well as indicate changes in "intensities." In his study of documentaries on television, Corner argues that concerns about bias and what he calls "journalistic rationalism" have kept music in documentary at a distance. The tendency of documentary films toward direct address and their claims to truth mean that "musical relations" are "more self-conscious" and "less intimate" in documentary films than in fiction films. Some documentary filmmakers worry that added music will be an "importer of unwelcome emotion and feeling" (Corner 2002, 357–358). In fact, throughout the 1960s, music in documentary films was permissible only if it was part of the film footage (Ruoff 1992, 226). Corner finds that the "lighter" the subject matter of documentaries, the greater allowances for music. Conversely, when documentaries consist of more serious subject matter, filmmakers scrutinize the potential effects of the music more critically (Corner 2002, 362). In some instances, then, musical interventions place the claim to realism of documentary films in a precarious position.

Corner's assessments of the role of music in documentary filmmaking translate fluidly to television news. Both documentaries and news programming claim to transmit unfettered reality, and an extradiegetic musical soundtrack can endanger that commitment to unadulterated realism. Corner's work implies that excessive use of music in the terrain of documentary filmmaking would diminish, if not eradicate, the gravity of the text if realism were its aim. The opposite is true in high-concept films, as Wyatt reminds us. In that context, music is sometimes used to overpower the narrative in a way that supports the simplified concept of the film. Both arguments coalesce in the area of commercial television news. The insights of Belton, Ellis, and Corner suggest the significance of sound in the genre of 24-hour television news programming.

## The Sounds of War

The most sonically significant moments in the first week of the 2003 war coverage include music and sound effects the networks used to accompany images— the extradiegetic sounds—and the naturally occurring sounds that emanated from the guests, studio journalists, and the war zone—the diegetic sounds. Both extradiegetic and diegetic sounds signified elements of the marketable concept. CNN and Fox News Channel used sound to remind viewers of the motivation and superiority of the U.S. military and its technology and to differentiate themselves from other networks for commercial purposes.

## *Music and Visuals in Title Sequences*

Whether music in high-concept films explicitly comments on narrative action or not, it reinforces the marketable concept. Music has a parallel role in television news, where it maintains ties to high concept. John Williams, who scored such high-concept films as *Jaws* (1975) and *Star Wars* (1977), also composed the theme to *NBC Nightly News* (Kirby 1999). Williams's success as a composer of Hollywood film scores stems from his ability to capture the marketable concept and articulate it in his work. Musical themes like the highly recognizable ones from *Jaws* and *Star Wars* carry a degree of pop cultural capital that translates into marketing potential. A well-written musical phrase (for example, the ominously deep tones of the *Jaws* theme) can carry as much weight as a striking image (the poster of *Jaws,* for instance, in which an enormous shark lunges up at an unsuspecting female swimmer). The same holds true for music on news programming.

In the television industry, networks sell audiences to advertisers. Viewers must be attracted to the program to complete the transaction, so networks use music to promote the program and the marketable concept. This makes music a valuable marketing tool. Not surprisingly, practitioners' ideas about the functions of music on television news do not differ substantially from the ideas of academic theorists. Former CBS News advertising executive Jeff Kreiner identifies music as a tool for grabbing the attention of distracted viewers, and composer Shelly Palmer argues that music is a technique for "branding the news" (Kirby 2003). Composer Bob Israel argues that music emotionally fortifies the content of the news (Engstrom 2003). These three purposes—capturing attention, making consumers aware of a product, and rousing emotional investment—capture the essence of high concept on television news.

The tension between journalistic values and musical intervention has existed in television news since its inception. In 1959, executive producer of *CBS Reports* Fred Friendly selected an Aaron Copland piece to accompany the news program (Engstrom 2003). Friendly's move was seen as bold because the news was perceived as a "sacrosanct" institution that rejected such obvious entertainment values. Richard Salant, head of CBS News, reversed Friendly's decision in 1961 and banned musical accompaniment for all CBS News programming. NBC took a different approach in 1963 when its new half-hour nightly news program concluded with a piece by Beethoven. Reuven Frank, the producer who decided to add the music, justified his move with the claim that "the teletype opening of Cronkite's show was used as music. It was no less artificial than the music we were using" (ibid.). That Frank felt he had to defend his choice reinforces Corner's notion that music is believed to compromise serious journalistic values.

Local news was more honest in its embrace of show-business values. They used marketing consultants to help them boost ratings and freely used music to open their programs (Stam 1983, 33). When network news executives imported local news personnel like Van Gordon Sauter in the 1980s, they also invited a market-oriented approach to television news. As a result, news producers began to position music carefully in their programming and pay great attention to the connotative meanings of the music they selected.

News producers want the music they select to clearly communicate something about their programs, and the process begins with the composers. Shelly Palmer, who composed MSNBC's theme music, was shown the network's sets and on-screen graphics to give him an understanding of "the look and the feel" of the network (Kirby 1999). News content varies little across commercial networks, so the look is a primary means of distinguishing one network from the other. Palmer's familiarity with the look of the network was central to his ability to capture its differentiation musically. Former NBC News president Lawrence Grossman argues that the development of news music relates directly to the development of the image. The stylistic effects of shooting on film—little editing and a slow visual pace—changed dramatically with developments in video technology, which allowed for more frenetically paced editing, graphics, and images. Grossman says that contemporary news music "matches and reflects the visual manipulation," and even though he does not connect the stylistic sensibility to the influence of entertainment values, his point underscores the institutionalization of music in television news (quoted in Engstrom 2003).

Since the late 1970s and 1980s, news theme music has been used explicitly to produce meaning. Bruce Brubaker, a faculty member at the Julliard School of Music, says that the ABC, NBC, and CBS news themes in use since the 1970s and 1980s are examples of the "fanfare"—military music featuring brass and percussion (quoted in Engstrom 2003). He explains that the fanfare is a genre specific to the military parade, which is itself a celebration of competition and conquest. The connotations of military music are not lost on composers or news producers. In wartime, those meanings intensify.

In 2003, all the mainstream news networks, including National Public Radio (NPR), embraced music as a way to differentiate their war coverage from that of the other networks. The message of war coverage music in 2003 depended on the intentions of composers and news producers. Composer Shelly Palmer touches on the complications of scoring war coverage with his comment, "War graphics and that deep, dark music trivializes and desensitizes you to what's going on" (quoted in Kirby 1999). At NPR, Jeffrey Freymann-Weyr altered a composition he had written for the coverage of the U.S. invasion of Afghanistan to accompany the network's coverage of the invasion of Iraq. The director of *All Things Considered*, Bob Boilen, described the highly praised music as "more compassionate music, more thoughtful without being sad, because no matter how you feel about this conflict, I think people feel compassion for the soldiers and the innocents" (quoted in Dobrin 2003). NPR executives pointed to the fact that they were not "saddled with graphics and live images that might dictate or influence the emotional tone of the music"—an advantage of operating in the medium of radio that led NPR to believe they could avoid the "jingoism or pacifism" exhibited by the television news networks (Von Rhein 2003). David Graupner, whose company developed *Juggernaut*, "an aggressive and at times overtly militaristic music package" for talk radio, took a different approach. Graupner, who finished the package in December 2002 in anticipation of a war or terrorist attack, was motivated by Fox News Channel. He confessed, "I would be a bald-faced liar if I said *Juggernaut* wasn't inspired by what you hear on that channel" (quoted in Engstrom 2003).

Von Rhein writes that CNN Radio used some of CNN's music in 2003, but it sometimes aired "more reflective piano music" at the producers' discretion (2003). All of CNN's music, however, was chosen from a commercially available pre-packaged set of themes (Dobrin 2003). Unlike Fox News Channel, whose producers commissioned music based on established visuals, CNN's managers matched their networks' graphics to the pre-packaged music. CNN's music functioned in three primary ways: as a reintroduction to coverage of the war after commercial breaks, in abbreviated form as a transitional device at half-hour updates, and in extended form under the vocal soundtrack of segments like "War Update" and sometimes "At This Hour." Although CNN maintained that the aim of the music was not "histrionics or drama," this claim seems a bit disingenuous; it explicitly used the music, particularly after commercial breaks, to reacquaint the viewer with the war narrative (Von Rhein 2003). The seriousness of the theme, made evident by the tempo and the melody, which was accompanied by a military-style snare drum, would not have been appropriate for coverage of a natural disaster or a presidential election.

While CNN's music did not connote patriotism outright, it did seem to yell, "We're at war!" as Peter Dobrin put it (2003). The main theme for CNN's *Strike on Iraq* title sequences (and overall coverage) was ominous yet fast, playing at about 120 beats per minute.[2] The brief piece began with a suspended cymbal roll and featured a snare drum playing under a melody in a minor key. The music was in standard $\frac{4}{4}$ time, but the bass line fell on beat one, on the eighth note after beat two, and then again on beat four. The effect was slightly destabilizing. There were no subtleties in volume; the entire piece was played loudly. The loud volume mixed with the destabilized beat to give the impression of a forward-moving determination without the stodginess of an actual military march. The piece ended on a minor note played by the chimes, giving a sense of a mission yet to be accomplished, suggesting that a block of information would complete the viewer's understanding of the mission in short order.

Titles and title sequences were another innovation that originated outside of the network news arena. Roone Arledge, whom ABC News hired away from *Wide World of Sports,* pioneered the use of special titles for themed news coverage—*America Held Hostage* was the title for coverage of the Iranian hostage crisis on *World News Tonight*. Titles implemented "splashy, elaborate openings" that predominate even in war coverage (Wittebols 2004, 77).

Although the first title sequence for CNN's *Strike on Iraq* coverage was forceful musically, it was less forceful visually. The iconography avoided any explicit reference to the United States, though the color scheme did hint at the red, white, and blue of the U.S. flag. In this instance, only part of the marketable concept came across. At the beginning of the sequence, a gold-colored representation of the country of Iraq in the center of the frame increased in size. Outside the borders of the country was a reddish hue with deep blue in the upper and lower left-hand corners of the frame. The capitalized words "Strike on Iraq" were barely discernable and appeared at the center of the frame intermittently. Meanwhile, the golden Iraq continued to grow larger, with rays shooting out from behind the outline as well as from the letters in "Strike on Iraq." The rays became golden and blinding white, turning to red with flecks of gold. At various locations within the frame, what appeared to be random white letters in varying sizes materialized. As an Iraqi flag entered the frame from the left side of the screen, the words "Strike" and "on" became more apparent. Frame by frame, slight changes occurred until the word "Iraq" became visible at the bottom of the screen. Finally, "Strike on Iraq" found its place in the center of the frame,

---

2. I differentiate CNN's *Strike on Iraq* coverage from its *War in Iraq* coverage, each of which had its own theme song. CNN switched to the *War in Iraq* logo and theme music after the first days of the invasion. That theme was slow and featured low brass.

flanked on each side by the Iraqi flag. The outlines of the letters appeared to remove themselves and come toward the viewer, finally disappearing in a flash of white light. The background was reddish-gold, and gold flickered across the frame as the sequence ended.

The imagery in a second version of CNN's title sequence was arguably more explicit for its *Strike on Iraq* programming; here, the indexical signs of the U.S. military graphically invaded the iconography of Iraq. In this alternate version of the "Strike on Iraq" title sequence, the golden Iraq moved from screen left to the center of the frame. Reddish-gold beams shot out from behind it on either side. The golden color inside the Iraq graphic gave way to video footage of a U.S. aircraft carrier. "Strike on Iraq" appeared erratically on the screen, as in the first version. The entire image grew in size as the camera seemed to zoom into Iraq with the aircraft carrier literally inside it. The aircraft carrier image faded as the Iraq graphic again moved to screen left. The "Strike on Iraq" logo entered the frame with the Iraqi flag on either side of the word "on." Shots of a jet landing and U.S. troops marching replaced the image of the aircraft carrier, but as CNN reused this sequence, the images varied.

CNN's title sequences denoted action taken against Iraq, but the first sequence did not explicitly name the aggressor. In the first sequence, the images of Iraq and the Iraqi flag represented the official government and geography of the target. The only hint of U.S. involvement came in the form of the colors red and blue. The second version of the title sequence contained more aggressive and overt references to U.S. technology and the marketable concept, but the visual references were still fairly subtle.

Fox News Channel's war coverage theme music and title sequence were markedly different from CNN's. The composer, Richard O'Brien chose music that jolted him because "hearing such a high sound will make anyone in a room instinctively turn around and look" (quoted in Engstrom 2003). In his words, the Fox News Channel's original theme was like "Metallica rehearsing Wagner, the guitar chords rising over thudding drums. It seemed ready-made for *Apocalypse Now*. . . . But we wanted the music to say, 'Something big is coming this way'" (2003). The war theme was in keeping with Fox News Channel's policy of "keep[ing] the sound and look younger and hipper" than that of the competition. O'Brien places Fox News Channel's musical choices in opposition to the "other networks" that "always go for that John Williams, big, grand music." In contrast, Fox News Channel's music "is always pointedly more aggressive." He deepens the divide between Fox News Channel and the competition, calling the other networks' management "a bunch of arrogant journalists" whose "style is so anesthetized" (O'Brien quoted in Dobrin 2003). Indeed, FNC's custom theme was less subdued than CNN's, but O'Brien did call for its alteration before it aired. He claims the first version of the theme was "too shrill, too rock 'n' roll" (Dobrin 2003).

Like the *Strike on Iraq* theme, the *Operation Iraqi Freedom* theme was played loudly in standard $\frac{4}{4}$ time. Unlike CNN's theme, Fox News Channel's music maintained a steady drumbeat on all four beats per measure with tom-toms and a low-pitched snare drum playing most of the sixteenth notes for every beat. The melody, played by horns (or keyboards), strings, and electric guitar, was less ominous than the music for *Strike on Iraq* and much more motivational, as if it accompanied an action sequence. Two sound effects—jets swooshing by and an eagle's cry—combined with the instrumentation to infuse the sequence with a militant brand of patriotism. O'Brien states that the addition of tom-toms that sounded like "war drums" added an extra sense of "urgency" (quoted in Dobrin 2003).

The *Operation Iraqi Freedom* theme and title sequence invoked the market-able concept as well as the identity of the entire network. Fox News Channel took its war coverage title from the Pentagon, though it appeared in the title sequence to be a subtitle under another logo (and another official title), "War on Terror." The title *Operation Iraqi Freedom* denoted not hostile action but liberation. The two agents in the Fox News Channel title sequence were the United States, represented by a fighter jet and the American eagle, and Fox News Channel, whose aggressive link to the Bush administration's "War on Terror" made the channel an active participant in that war.

In the title sequence, a large, incomplete Fox logo rotated to the top of the screen over a blue, black, and white background. A fighter jet appeared and flew from screen left to screen right. As the Fox logo continued to rotate, another jet appeared from behind it on screen left, turned forward, and fired two shots. As it flew, the jet morphed into an American eagle. The eagle continued to fly forward and then toward the lower left-hand part of the screen. The Fox logo continued to rotate, and crosshairs appeared over the "o" in "Fox." "War on Ter-ror" appeared over the crosshairs, and "Operation Iraqi Freedom" subsequently replaced "War on Terror."

In another Fox News title sequence, a representative of Iraq finally ap-peared. The sequence began with a U.S. flag filling the entire frame. Thick white vertical bars entered the frame and moved to screen right. Footage of Saddam Hussein materialized on screen left. As the white bars moved off the screen, a U.S. jet appeared on screen right. The U.S. jet then disappeared, and a U.S. tank appeared on screen left as the image of Hussein dissolved. The tank, too, disappeared. A white horizontal bar appeared mid-frame and morphed into the text "The Cost of Freedom." The title moved to the top portion of the frame to make room for "War on Terror." The indexical signs of both Hussein and the U.S. military (but not U.S. leaders) finally affirmed the presence of the war's two primary opposing forces, but the sequence demonstrated an obvious affinity for the U.S. side.

Along with its strongly thematic title sequence, Fox News Channel included strategically placed sound effects throughout its war coverage. These tactics are examples of the "italicizing" function of sound, which, according to Altman, pulls viewers back to the television set when they have wandered away or are generally distracted (1986, 45). Fox News used sound effects during quick transitions during which the logo of the channel's war coverage would appear for a brief moment. The traditional chime for the "Fox News Alert" was played during the war coverage, signaling that viewers needed to direct their attention to breaking news. And during the network's *Operation Iraqi Freedom* graphic sequence, a definitive boom accompanied the appearance of the war coverage logo on the screen. Whereas CNN used primarily music to italicize segments, Fox News Channel fully utilized multiple sonic elements to retain its audience.

The theme music and title sequences that CNN and Fox News Channel used maintained two realms of signification that supported one another connotatively. While the images explained the narrative and the marketable concept to varying degrees in different versions of the networks' title sequences, the earnest, almost military-sounding music provided fitting accompaniment. Each network used elements of entertainment—music, sound effects, graphics, and animation—in its title sequences to articulate how it was different from the other network. Theme music and sonic embellishments transcended the role of news dissemination and moved the entire enterprise further into the terrain of entertainment. Such evidence abundantly indicates that television news is heavily marked by the entertainment industry.

### Anticipation and Detonation

Those who want to be considered serious television journalists stand firmly against tampering with diegetic sounds. Jill Rosen (2002) writes of an incident in Las Vegas, where local reporters showed surveillance tape of a shooting with dubbed-in sounds of gunshots and ambient casino noise. Deborah Potter (2004) blames sensationalism and the fact that news producers have the technology to create such effects for incidents like this. Rosen criticizes the practice and offers ironic suggestions for expanding it into all types of news stories. For war coverage, she recommends using footage from old war movies, removing their scores, and adding the "leftover gunshots from the casino" (Rosen 2002). Diegetic sounds in television news are the keys to journalistic fidelity.

Diegetic sound not only verifies the image, it also accounts for most of how the news communicates—through language. Coverage of the war at CNN and Fox News Network supplied five major types of diegetic sound that gave even the on-air personalities pause. Sirens, calls to prayer, explosions, weapons deployments, and silence each contributed to the narrative and supported the marketable concept in unique ways. Diegetic sounds anchored the narrative and the ideology to unfolding events, effectively legitimizing the narrative by making it *sound* real. Meanwhile, news personnel used these sounds to frame events in terms of high technology rather than human suffering.

Air-raid sirens, heard primarily in Kuwait, were of special consequence to CNN's and Fox News Channel's correspondents based in Kuwait City. Not only did the loud, unsettling sirens indicate potential danger for the reporters, they also signaled forthcoming Iraqi aggression, which propelled the narrative. On several occasions, CNN and Fox News Channel correspondents in Kuwait invited everyone to listen, even teaching viewers how to distinguish between the different sirens. Contributing to the eeriness of the sirens were the muffled voices of the journalists wearing gas masks. The sounds interacted with the images in unsettling ways. The static shot of CNN's Bill Hemmer, for instance, standing calmly on a hotel balcony overlooking a section of Kuwait City contrasted sharply with the action that the air-raid sirens signified. Similarly, reporters from both networks carried on with their work as loudspeakers issued calls to prayer throughout Baghdad. In the first days of the invasion, journalists explained the sounds for the television audience, but as the coverage continued over a period of days, the prayers became background noise and faintly heard examples of local color.

Most of the attention to diegetic sound on CNN and Fox News Channel focused on the noises of the bombings and firefights. The "shock and awe" campaign provided many opportunities for network news personnel to marvel at the sounds of the explosions and, by implication, the sounds of properly functioning U.S. technology. At approximately 6 AM Baghdad time on March 19, Aaron Brown invited the audience to listen to "the sounds of Baghdad this morning," referring to the normal sounds of traffic and the alarming sounds of anti-aircraft artillery fire (CNN March 19). Once the bombings began, Wolf Blitzer called the sounds of the explosions "nerve wracking," and Dr. Sanjay Gupta described hearing "thuds and booms" overhead while he took cover in a bunker (CNN March 20). Thomas McInerney identified which types of missiles had been deployed based on the sounds of their explosions (Fox News Channel March 21). Embedded reporter Rick Leventhal even held his satellite phone out on two occasions during a firefight so that the audience could hear what he described as "thumps" (Fox News Channel March 20).

Furthermore, the stationary cameras placed in various locations around Baghdad often did not capture images of explosions, so for viewers, the sounds the camera microphones captured were the only evidence of the bombings. In cinema, off-screen sound peaks the interest of the viewer (and the camera) because it has no discernable source. Eventually, the camera finds the source. In television, Altman argues, the "spectator/auditor" assumes the role of the camera when searching for the source of the sound. "By raising my eyes and glancing toward the screen," Altman writes, "I discover the sound source on my own, thus experiencing the wholeness it implies" (1986, 46–47). When neither the camera nor the viewers could locate the source of the explosions on screen, news personnel simply emphasized the value of the sounds. For example, Shepard Smith described hearing two explosions in Baghdad on the night of March 21 that he could not see. In a search for proof of the continuing air campaign, Smith invited the audience to "listen in and see if we can hear anything" (Fox News Channel March 21). On March 23, when major explosions resumed in Baghdad, Lou Dobbs noted that you could not see the explosions, "but you can hear them" (ibid.). In the absence of images, sonic traces verified the existence of the explosions.

The sheer power of the sounds—sometimes so loud that they drowned out any verbal communication—helped place the spectacle of the military apparatus at the forefront of the coverage. Along with the sounds of bombs and other types of exploding ordinance were the sounds of weaponry and aircraft. Planes elicited great notice on the networks. Fox News Channel personnel made a point of stressing throughout March 21 and March 22 that every strike aircraft in the U.S. arsenal was used during "shock and awe," prompting Neil Cavuto to comment, "We are strutting our best stuff, aren't we?" (Fox News Channel March 22). That pride was typical of Fox journalists' reactions to U.S. military technology. Both Shepard Smith and Bret Baier asked the audience to listen in to the sounds of aircraft taking off from the USS *Constellation* (Fox News Channel March 21 and 22). The sounds of jets taking off and landing often were so loud that Bob Franken's words became inaudible (Franken frequently reported from an unidentified airbase). Likewise, embedded reporters based on aircraft carriers who reported during peak traffic transmitted little more than the overwhelming sounds of jets speeding away. These sounds, whether journalists commented on them or not, were testaments to the technological power of the United States.

Perhaps the most conspicuous response to the aural events was the rare decision of news anchors to allow the sounds to fill the airwaves without any commentary. On Fox News Channel, a Sky News anchor explained that network's editorial decision not to talk over the images of "shock and awe"; the British anchors at Sky News sat silently for ten minutes during the bombings (Fox News Channel March 21). CNN and Fox News Channel news personnel did not follow this policy; allowing the sounds to dominate without commentary would have liberated the event from the narrative. Corner writes, "Silence presents the possibility of an embarrassing insufficiency of meaningfulness and a more embarrassing uncertainty about whether this insufficiency is essentially in the work or in the viewer" (2002, 360). In the case of the war coverage, where the "work" was the shaping of the image through visual and verbal discourse, silence in the studio while airing images and sounds of bombs exploding might have revealed the usual commentary to be blatantly irrelevant or at least insufficiently meaningful. Any decision to stay silent on CNN and Fox News Channel was spontaneous and very short-lived. Wolf Blitzer was one of the few proponents of keeping quiet; he halted his reports briefly on several occasions to allow for a complete dedication to the sounds of anti-aircraft artillery fire or explosions (CNN March 21–23). Bret Baier and Shepard Smith each requested quiet for one minute to listen to the events as they unfolded, but both times other on-air personalities broke the silence within ten seconds (Fox News Channel March 21). It was as though pure nonverbal diegetic sound was the equivalent of a black screen; it was dangerous and intolerable.

CNN's and Fox News Channel's attention to sounds and certainly the sounds themselves raised several issues about the role of sound on television. First, the network journalists' deference to sounds when cameras could not capture images challenge Belton's claim that sounds depend on images for their realism. When this happened, sound was the only available evidence of reality. The "wholeness" that might have existed if there had been images to match the sounds was beside the point at CNN and Fox. The aural proof of the continuing U.S. bombing campaign was enough proof for the two networks, and their narrative did not waver. The sounds functioned as evidence of superior technology. Secondly, the news personnel's fascination with the sounds recontextualized the enormity of the significance of the sounds. Before-and-after coverage trained on the sounds of launches as weapons left their bases and on the detonations when they found their targets amounted not only to a fetishization of the technology but also to an outright denial of the inevitable human suffering. The network journalists' faith in precision weaponry erased the topic of the dead and dying Iraqis from their coverage.

## Sights of the War Narrative

In a very basic visual sense, CNN and Fox News Channel created two different identities, or brands, for their war coverage using similar tactics. Though both networks displayed visual flair, each crafted that flair differently. I have already discussed how Fox News Channel prided itself on looking and sounding young and hip; its visuals reflected its overall mission. Each network's visual aspects—the news frames, graphics, embedded footage, representations of the war apparatus, and use of still photography—carried its own high-concept characteristics.

Some typical examples of the visual qualities of high concept are "extreme backlighting," "minimal color scheme," "predominance of reflected images," and "a tendency toward settings of high technology and industrial design" (Wyatt 1994, 17). High-concept films privilege "sleek, modern environments mirroring the post-industrial age through austere and reflective surfaces" (30). At times, these visuals become so overpowering that they draw too much attention to themselves and detract from the narrative. In this way, they can be another example of excess attributable to a lack of authorship in the traditional cinematic sense. For instance, any discussion of stylistic excess in classical Hollywood cinema must include a mention of—if not a preoccupation with—the Douglas Sirk melodramas of the 1950s. Wyatt maintains that the melodramatic plots of these types of films reconcile the tendency toward excess. In high-concept films, this reconciliation does not take place because the excess lacks a fundamental link to authorship or "personal vision." High concept is self-conscious, but the director of the film is not the creator of the style. The excess is present solely at the behest of the marketplace, which Wyatt claims is "clearly the author" (34). The same "industrial expressivity" that characterizes the content and marketing of high-concept films is evident in television news (61). From high-tech stylization to striking images easily extracted for marketing purposes and finally to sheer excess, the visuals at CNN and Fox News Channel functioned commercially and narratively with enduring faithfulness to the marketable concept.

### Graphic Differentiation

The visuals on CNN and Fox News Channel were similar in basic composition but different stylistically. Title sequences were one means of differentiation. CNN's title sequences were characterized by constant motion, randomly appearing script, and brilliant color. The color scheme and the use of the Iraqi flag were the most overtly political facets of the sequence, with the logo and the theme music emphasizing a general forcefulness. Fox News Channel's title sequence also was characterized by constant motion, bold text, and intense color. Unlike CNN's initial title sequence, though, Fox News Channel's title sequence was more overtly political in its iconography.

The networks' graphically sophisticated title sequences selectively and subjectively revealed information about the invasion and about the narrative. Each sequence conveyed the marketable concept according to the style of the network. The networks' visual approaches to the conflict also reflected their overall coverage, although I found that CNN was more aggressive in its representations of the invasion and the U.S. military. Through an examination of the visuals, we can recognize a level of commitment to the war narrative at CNN that was stronger and more overt than what CNN's on-air personnel conveyed in their commentary.

As visually, politically, and ideologically busy as the war-oriented title sequences were at the two networks, the everyday news frames were even more loaded with images and stylized information. Both CNN and Fox News Channel maintained the same basic types of frames.[3] They consisted of a full-frame image; split screens; two unequally sized boxes of footage placed side by side; images of maps alongside still photos of reporters; videos of reporters alongside larger boxes of footage; and videophone images of embedded reporters alongside a map of their approximate locations. Text crawls or news tickers were mainstays at the bottom of the screen, as were the networks' logos on the lower left corner of the frame. At CNN, slightly above and to the right of the network logo was the text "Strike on Iraq." At Fox News Channel, a graphic of the U.S. flag accompanied the network logo; it had occupied that space since September 11, 2001. Both networks used a banner above the text crawl that displayed various titles and information describing the frame's images. On CNN, the titles were "Breaking News," "Strike on Iraq," or whichever title was appropriate for the report that was airing. On Fox News Channel, the titles were "Operation Iraqi Freedom," "War Alert," or others.

CNN and Fox News Channel each used distinctive frames that expanded on the basic designs described above. CNN's frame for the reports of embedded

---

3. I am using the phrase "news frame" to refer to the arrangements of the elements on the screen during a news broadcast.

journalists consisted of the typical graphics on the bottom of the frame and more elaborate graphics on the top half. A still photo or a videophone image of the embedded journalist appeared in a box to the left of center. Above and to the right of the image box was the logo of the military unit to which the journalist was assigned. Below the logo was text that revealed the name of the reporter and below that was the name of the military unit. This basic design was reversed on Fox News Channel, where the photo of the reporter was on the right, and his/her information appeared below the image. On the left side of the frame was the "War on Terror" logo over a U.S. flag. Underneath that main title was the logo and name of the military unit. CNN and Fox News Channel both used visual and textual descriptions of the military units of embedded journalists, but Fox News Channel's addition of the large U.S. flag connoted a degree of nationalism that CNN's frame did not.

For the most part, the remainder of the frames demonstrated a proclivity for stylish and complex graphics. (Appendix A provides detailed information about the graphic frames at both networks.) CNN featured at least three different frames using image boxes that displayed feeds from various locations. When it used more boxes, usually the visuals were more significant. For example, a frame with four equally sized boxes appeared most often during the bombings to show explosions that had been caught on four different cameras in Baghdad. Other frames on CNN were thematic and provided information regarding updates, "latest developments," upcoming news, "war recaps," "battle scenes," and embedded journalists. Fox News Channel used two-, three-, and four-box frames to perform the same function as similar frames did at CNN: to maximize certain aspects of the war coverage in a visually appealing and high-tech (if cluttered) fashion. Fox also used two main thematic frames: one for general updates and the second for issues pertaining to the "war on terror."

Whether they were filled with political iconography or not, news frames at CNN and Fox News maximized space to provide viewers with an abundance of textual and visual information. The look of the two networks was technologically sophisticated and at times quite complex. On several occasions, CNN and Fox News Channel personnel had to interrupt their discussions to clarify which images were in which boxes and which labels of "Live" and "Earlier" were incorrect. Moments like those exemplified the assessment of the Project for Excellence in Journalism that the networks were "too infatuated" with their own technological prowess (2003).

Fox News Channel's look was edgier and more frenetic than CNN's, but both networks projected a look that aligned their overall styles with sophisticated technology while preserving the marketable concept of superior U.S.

military technology. Although the concept remained intact, the stylization of the frames contributed little to the narrative. Aside from the narratively relevant political iconography in Fox News Channel's logo and the language of the titles, the look amounted to moments of excess—pure stylistic indulgence that did little to advance the story.

### Visualizing the Spectacle of the War Apparatus

Both networks expanded on these basic stylistic elements and excess in their representations of the war apparatus. CNN and Fox News Channel journalists maintained a fascination with military hardware that surfaced most conspicuously in visual representations of military assets, targets, and maneuvers. These representations contributed to the overall look of the two networks because they required even more visual embellishment than the typical news frames discussed above. Moreover, the way that CNN and Fox News Channel provided visual information about the U.S. military was central to the networks' construction of the military as superior, courageous, and heroic. The machinery of war was likewise the subject of awe and discursive celebration. In short, at both networks, the weaponry was treated as a spectacle. Giroux writes that "representations of the bodies of mangled civilians, children, and others who are not soldiers can be found on Al Jazeera and other media sites . . . but are rarely seen in the mainstream media in the United States" (2006b, 4), and this was certainly true at CNN and Fox News. On-air personalities regarded the images of bombings, planes, missiles, and mechanized infantry with wonder. They left it to other news outlets to provide images of casualties; both networks virtually expunged the human element from their discussions of military hardware and strategy. This type of coverage of weapons, maneuvers, and targets shaped the view that the U.S. military was pristine and courageous, a perspective that harmonized with the audio and visual information the networks chose to air.

CNN's discussion of wartime images ranged from the issue of constraints on the networks—Wolf Blitzer conceded that CNN was "limited in the actual pictures" the network could show—to effects on viewers—guest Alex Jones warned of a potential "coarsening"—and finally to strategic usefulness—Blitzer noted that the images were "confirmation for the military that they have hit targets" (CNN March 20 and 23). The images also incited on-air journalists to attempt to brush off the kind of criticism that remained from the 1991 Persian Gulf War (that the war coverage was excessively stylized, etc.). Aaron Brown repeatedly assured viewers that the coverage was "not a television program" and "not a video game" (CNN March 20 and 22). Likewise, Tony Snow and David Hunt asserted that the images were not a video game, while Hunt and Brit Hume reminded viewers that they were not watching a movie (Fox News Channel March 22 and 23). In a rare departure, Alan Colmes stated that the coverage was "almost like a video game at times" (Fox News Channel March 24). For the most part, CNN and Fox News Channel news personnel were cognizant of the criticism that their war coverage diminished or exploited the hazards of war. At the same time, they actively fetishized the images they displayed.

Images of bombings elicited a great deal of excited description from news personnel apparently starved for visually appealing events. CNN's Wolf Blitzer described the bombings in Baghdad on the first night of hostilities as "spectacular," and Aaron Brown described the visual effects of the bombs as "orange fire . . . lighting up the sky" (CNN March 23). Fox's Shepard Smith described the images of Tomahawk missile launches off the USS *Constellation* as a "show of force and power" and a "fascinating sight" (Fox News Channel March 21). Rita Cosby called the images of the "shock and awe" bombing campaign "amazing pictures" (Fox News Channel March 22). Chris Klein observed the bombings of Mosul and Kirkuk through night-vision goggles and described them as a "constant pulse of incandescent lights." A short time later, he said that the bombing was an "absolutely spectacular sight" but acknowledged that "certainly a great deal of death and destruction [was] raining down . . . in those two cities" (Fox News Channel March 21).

Studio personnel at both networks breathlessly offered their observations and sometimes eloquent descriptions of images of bombs being launched and dropped. Equally important was their enthusiastic anticipation of these images—an anticipation that solidified their desire for something to happen that was visually stunning and not just operationally (or narratively) significant. This attention to visuals invokes not only the stylistic demands of high concept but also the commercial value of arresting images. Explosions draw viewers in, and that fact was not lost on CNN or Fox News Channel personnel. They celebrated the images of the war apparatus unconditionally and gave little or no consideration to the lethal consequences of the bombs, missiles, and other artillery.

Both networks used framed graphics that resembled baseball cards to display U.S. weapons. These images celebrated the unparalleled arsenal of the U.S. military. Each graphic consisted of a computer-animated weapon and text that provided basic facts and statistics about the weapon. Overloaded with data and computer-animated images, these graphics represented an excessive style that illustrated the networks' fascination with and celebration of U.S. weaponry.

At CNN, these visual presentations of U.S. weapons were particularly important indicators of the network's enthusiasm for the marketable concept and war narrative. Much of the information on that network that conveyed the message that the U.S. military was technologically superior did so with visual images rather than with verbal commentary. The following description of CNN's graphic for the F-117A Nighthawk Stealth Fighter is a typical example of how the network presented visual information about U.S. weaponry. The frame consisted of a dark blue background and a bright golden outline in the shape of Iraq. In the foreground was the animated dark-gray figure of an F-117A Nighthawk plane. The F-117A flew from the upper right-hand corner of the frame into a more prominent position over the outline of Iraq. The flight gave the illusion of depth because the plane grew larger as it moved into the frame. A gold-outlined box depicting the U.S. flag emerged from the upper right-hand corner of the frame. Underneath the flag was the text "United States," and beside the box text appeared that read "F117A NIGHTHAWK STEALTH FIGHTER." The plane flew under the golden box to reach the outline of Iraq. As the plane hovered, statistics appeared on screen left. These included speed, range, weapons, and function. Each graphic for topics related to the Iraq invasion followed this general design, including the graphics for Saddam Hussein, the Iraqi Air Force, and the U.S. Airborne Division.

Fox News Channel's weapons graphics had a diagonally positioned Fox logo in the background in large bold letters that made the word nearly illegible. The color scheme was heavily blue and green. The network's framed graphics were superimposed over the logo in a self-consciously computerized style. In the graphic for the Tomahawk cruise missile, the first thing to appear on screen was the rotating missile. After the missile moved to screen left, the rest of the frame materialized. Screen right contained data about the missile: its description, capabilities, and background. As at CNN, all of Fox News Channel's framed graphics on topics related to the war in Iraq—including those displaying figures like Tariq Aziz, the Hussein sons, and U.S. soldiers—followed this design. Fox News also had a live-action graphic for every branch of the armed services, an affirmation of Paul Virilio's assertion that soldiers are not humans but "a primary material, an instrument that exists to be employed" (2002, 76). In each framed graphic, a representative from a certain branch of the military modeled his equipment over the sounds of a military-style cadence on a snare drum. In the graphic for the U.S. Marines, for example, two soldiers simulated hand-to-hand combat after a single soldier modeled the equipment.

The graphics of U.S. weapons at both networks conveyed the view that the coalition had the superior stockpile. Neither network devoted much time to explorations of Iraqi weaponry. Only the cards devoted to the Hussein family, Tariq Aziz, Scud missiles, and Ricin toxin contributed to that part of the narrative. The hyperstylized construction of the framed graphics created another moment of excess. The narrative invariably stalled as CNN and Fox News Channel paraded the superior arsenal of a wealthy nation. Those displays were pure spectacle that used an excessively stylized mode of delivery that obscured the issue of how many people the weapons would kill.

The weapons fetish was also evident in the computer-animated vignettes of the arsenal in action that both networks periodically used. These visually appealing animated sequences demonstrated an idealized vision of the weapons' operations in a way that both supported the war narrative and advertised the weapons. Some of the sequences on CNN included the flight of a cruise missile, a Stealth jet and guided bomb (also known as a "bunker buster") scenario, troop progression, the capture of an airfield, an armored formation, an Apache helicopter in flight, and a B-52 launching a conventional air-launched cruise missile. The animated flight of a cruise missile aired early on March 20 before the "shock and awe" campaign and offered a casualty-free, sanitized view of what would happen on a large-scale bombing run.

The animated sequence of the Stealth jet and the bunker-buster bomb showed viewers in retrospect what should have happened during the "decapitation" strikes on March 19. The bunker buster is equipped with a 500-pound warhead designed to detonate once the bomb has penetrated a concrete structure. The animation conveyed this scenario in action (see appendix B). The most significant part of this animated scene was not the illustration of the bomb's function but the explosions that it caused. In the sequence, a bunker buster penetrates three levels of a warehouse. The explosion on the upper level of the warehouse produces black smoke, while the smoke from other explosions is green. The animation took for granted the "fact" that Iraq was storing chemical weapons. Another animated sequence that illustrated the operation of the Patriot missile also presumed that Iraq was using Scud missiles, even though the Pentagon later denied the presence of Scuds. The computer-animated vignettes on CNN fetishized the operation of the weapons and obscured their effects on the lives of Iraqi civilians, in the process supporting the war narrative that the United States was in Iraq to liberate its citizens rather than harm them.

Fox News Channel's animated sequences clung to the war narrative as fervently as CNN's did. Some of the network's animated sequences included simulations of Patriot missiles intercepting Scud missiles, battles between U.S. and Iraqi tanks, and the maneuvers of the U.S. Navy fleet. To contextualize the Patriot-Scud intercept sequence, I must add that military officials claimed and news personnel repeatedly stated that the new generation of Patriot missile (the PAC-3) was an improvement on the type used in 1991 in part because of its ability to hit a missile or an aircraft instead of simply detonating near it. The Scud-Patriot duels had been a huge story in the early days of the 1991 Persian Gulf War. Philip Taylor sums up the intrigue of the missiles:

> The success of the American Patriot missiles in intercepting the Scuds provided, in microcosm, a televisual symbol of the conflict as a whole. It was a technological duel representing good against evil: the defensive Patriots against the offensive Scuds, the one protecting innocent women and children against indiscriminate attack, the other terrifying in their unpredictable and brutal nature. (1998, 70)

The drama of the Gulf War duels gave way to reports that the Pentagon's claims that Iraqi Scud launchers had been destroyed by Patriot missiles were misleading, to say the very least. Gradually, more information came to light about how often the Patriots missed their targets; an investigation in 1992 by the House Government Operations Subcommittee on Legislation and National Security revealed that Patriots had hit only a few of the Scud missiles the Iraqis had launched (Frontline 1996). The danger of falling debris also sullied the reputation of the much-lauded U.S. anti-missile system. In 2003, CNN's Bill Hemmer was vocal about the poor accuracy of Patriot missiles in 1991 and about the "fact" that no Scud launchers had been destroyed (CNN March 20). By March 21, 2003, on-air personalities felt that the new Patriots had improved on their past performance. Even after news broke early on March 23 that a Patriot missile had mistakenly shot down a Royal Air Force Tornado jet, both CNN and Fox News Channel personnel continued to praise the success of the Patriots. In response to the news, Fox News Channel's Bob Sellers dismissed the negative news by wondering aloud "how many lives they've [the Patriots] saved by working properly" (Fox News Channel March 23).

Computer-animated Patriot-Scud sequences reflected the optimism of Fox News Channel. Like the CNN animation of the bunker buster, Fox News Channel's Patriot-Scud animation presumed that Iraq possessed illegal conventional weapons (see appendix B).[4] The animation posited a scenario in which the Iraqis

---

4. UN Security Council Resolution 687 (1991) mandated that Iraq destroy its ballistic missiles whose range exceeded 150 kilometers. Scud missiles, which had the capacity to carry chemical, biological, and nuclear warheads, were therefore deemed illegal (Katzman 2003).

used their Scuds and mobile missile launchers. (The U.S. military never found any evidence that the Iraqis had mobile missile launchers.)

CNN and Fox News Channel also used satellite imagery provided by Digital Globe. CNN used satellite images of Baghdad to pinpoint past and potential targets. Fox News Channel digitally enhanced its satellite images, dubbing the finished product "Fox Fly Over." The satellite imagery sequences were another example of the networks' lopsided emphasis on military strategy. Like the other graphics both networks used, these sequences were also moments of excess that fetishized technology while doing nothing to advance the narrative.

The approaches to visualizing the war apparatus at CNN and Fox News wavered between narrative relevance and spectacular excess. Weapons graphics, computer-animated sequences, and satellite imagery advanced the marketable concept and failed to point out the consequences these weapons would have for human beings. Although spectacle and fetish sanitized the networks' coverage of the deployment of U.S. weapons, the computer-animated sequences never interfered with the narrative. The framed graphics and satellite imagery did interfere; they sometimes even stopped the coverage of the war in its tracks. In all of these approaches to conveying visual information, an uncritical appreciation of the U.S. military machine and its strategies was the dominant approach.

### Images from Embedded Reporters

A crew member with embedded reporter Martin Savidge summed up the prevailing attitude toward embedded footage when, during a series of explosions, he was heard to exclaim, "Fuck, this is good stuff!" (CNN March 22). Fox News Channel's Bill Cowan noted that U.S. soldiers in Iraq would be able to watch this footage with their grandchildren and expressed his regret that he did not have footage of his fighting days in Vietnam. Later, he repeated his lament that he had "nothing to show" for his time in Southeast Asia (Fox News Channel March 21). The footage produced by embedded news crews was interpreted by the networks to be factual documentary footage that displayed the resources of the military in a visually exciting way—it was the twenty-first-century spectacle as Giroux envisions it. Discussing the heightened interplay between images and power, he writes,

> Not only have these new mass and image-based media . . . revolutionized the relationship between the specificity of an event and its public display by making events accessible to a global audience; they have also ushered in a new regime of the spectacle in which screen culture and visual politics create spectacular events just as much as they record them. (2006a, 20)

The convergence of the military's narratives and the networks' narratives endowed the footage with the power of justification (particularly since the images obscured the war's lethal consequences). Although the look of the footage was neither sleek nor stylized, it advanced the networks' aim of high-tech visual appeal—an aim made evident by their in-house graphic embellishments.

In the first five days of the invasion, video satellite linkups were rare, and much of the embedded footage came in the form of satellite images via phone transmission. The grainy quality of these images made the experience of the military and the reporters embedded with it seem dustier, harsher, and thus more real. Green-tinted night-vision footage of firefights compounded that look by making everything appear shadowy and insidious. The response of studio news personnel to this footage was one of sheer amazement, which made the footage more about spectatorial pleasure than about the hardships of war.

CNN's Walter Rodgers and Fox News Channel's Greg Kelly, who were both embedded with the 3rd Infantry, transmitted similar images to their home networks, to the delight of their colleagues in the studio. They documented the push into Iraq, and for hours the audience stared at the rears of military vehicles as the reporters rode in their Hummers and tried to see through the dust. Aaron Brown called Rodgers's footage "incredible" and "amazing," at one point exclaiming, "Oh, look at that!" (CNN March 20 and 21). Rodgers called the image a "lovely picture" and said that "sadly," the viewers could not get a sense of the sheer number of tanks participating in the convoy (CNN March 21). Anderson Cooper admitted that he was "transfixed" by Rodgers's images, Lou Dobbs could not "get enough" of them, and Bill Hemmer called the images "fascinating" (ibid.). Kelly's images generated the same type of response on Fox News Channel. Brian Wilson called them "astonishing" and Gregg Jarrett described them as "truly remarkable" (Fox News Channel March 21). Rita Cosby, Bill Cowan, and David Hunt described the shots as "amazing," "dramatic," and "remarkable," respectively (Fox News Channel March 22). Firefights and footage of aircraft on carriers provoked similar responses at the two networks.

On-air journalists praised the embedded footage as a "technological miracle," and even Aaron Brown—no stranger to satellite technology—expressed how "amazing" it was for him to watch the images while sitting in Atlanta (CNN March 21). Despite the inferior quality of the images transmitted by the satellite phones, technology triumphed.

CNN and Fox News Channel juxtaposed the quality of those "real" images with the eloquence of still photography. The embedded and unilateral print journalists (those who entered the war zone without military permission) in Iraq generated photographs of the war, and both networks drew viewers' attention to still images from the battlefield at specific points in the coverage. By using photos, the networks attempted to create a stable meaning, a permanent "caption," as McAlister puts it. However, the historical specificity of the stills and our inevitable reevaluation of them both counter the meaning networks tried to impose with the captions. McAlister reflects on the intentions and interpretations of stills like those publicized in 1991:

> By asking the right questions of a photograph, we may discover what rendered it powerful at a particular place and time. . . . Photographers invite us to see them as transparent representations, as a kind of historical record. Yet, photographs are inevitably fragments. (2005, 268)

They are also signifiers of a level of gravity and restraint typically shunned by moving images on the news.

CNN used still photographs as a transition to half-hour updates, displaying the photos as a slow-moving montage with the *Strike on Iraq* theme playing over them. Aaron Brown called the photographs "moments of war quite literally frozen in time" (CNN March 21). The photos of soldiers pointing guns at or giving water to surrendering Iraqis were devoid of the frenetic movement and excited rhetoric that the embedded footage relied on. Their stillness and relative subtlety accorded them a degree of cultural prestige lacking in the video documentation.

Fox News Channel characterized its still photographs as evidence of the just mission. In his introduction to a series of stills that functioned as a transition at the end of the hour, Brit Hume described the photos as documentation of U.S. soldiers "doing the dangerous work of Operation Iraqi Freedom" in Iraq (Fox News Channel March 24). While the photographs appeared, no verbal commentary steered viewers toward any particular narrative or ideological direction. The music performed that duty as the photos were left alone, supposedly displaying reality as it unfolded. Accompanying the photos was an atonal piece of music that consisted primarily of different percussive sounds. The music was somber, austere, and serious without the excited victorious military connotations of Fox's *Operation Iraqi Freedom* theme. Instead, the music was reflective with a trace of military influence. In the context of the war coverage, only still photography warranted such a subtle, introspective moment at Fox News Channel.

The photographs added an odd low-tech stillness to the frantic production of television news. They also permitted a greater deal of reflection than grainy images and quickly edited video could. However, the still photographs served the aims of high concept. Each image was striking, and some of them served the networks' purposes of promoting the official narrative of the war. As in 1991, when images of "soldiers framed against the backdrop of a blazing sunset replaced mangled bodies and bloodied wounds" (Giroux 2006b, 3), in 2003 the photos of the war promoted the same whitewashed version of reality. Hume's comment that the soldiers in the photos were "doing the dangerous work of Operation Iraqi Freedom" drew the images into the marketable concept while emphasizing the realism in the representation.

The idea that the footage provided by embedded news crews was a realistic version of the war was a constant refrain at both networks. The commentary networks provided about their access to embedded crews highlighted the role of technology in producing the material—which, though rough in appearance and sonically delayed, was a sensation in the minds of commentators in network studios. Network personnel in the United States marveled at the sights and sounds that embedded crews transmitted back to the states. At both networks, the fetish for all things military overshadowed the human realities of the invasion, and embedded reporters became part of the spectacle, enabling the war apparatus to monopolize screen time.

### Images from the "Other" Side

There was one moment when the physical toll of war received attention on 24-hour network news—when the networks received the footage of the soldiers of the 507th Maintenance Crew who had entered the Iraqi-held town of Nasiriyah, where they had been captured by the Iraqis. The video of the POWs, which was produced by Iraqi television, challenged the official war narrative and provoked outrage on the U.S. networks, where it posed a threat to the highly controlled flow of images. The POW video, which was aired by Al Jazeera and other networks around the world, featured a makeshift morgue with bodies strewn about, including some with fatal head wounds. It also included Iraqi television interviews with the captured soldiers.

CNN and Fox News Channel refused to air the POW video in its entirety and stated their reasons for this decision regularly. They chose instead to air one still image to represent the video footage. Each network used its own disclaimer to introduce whichever segment of the POW video it had decided was appropriate to air. CNN aired an almost indistinguishable still of what appeared to be bodies lying on the ground, which various on-air personalities prefaced by saying that the video had been "transmitted by Al Jazeera . . . was shot by state-run Iraqi television. These pictures and the interviews were disturbing. . . . CNN has decided not to show the video of those killed and will instead use this single image with no identifiable features" (CNN March 23). When CNN began to show a brief clip of the POWs, Aaron Brown rationalized the decision, consciously differentiating it from Al Jazeera's decision to run the entire video. He said that the brief segment was necessary to show the identity and status of each soldier (CNN March 24). The image Fox News chose to broadcast was preceded by the reassurance that "considerable reflection and consultation" had informed the decision (Fox News Channel March 23). Fox also assured viewers that the image did not violate Pentagon rules because it revealed no identifiable features. CNN and Fox News Channel conceded that the tape was newsworthy, but with every disclaimer they stressed their disapproval of Al Jazeera by underscoring how tastefully and responsibly they had chosen to use the POW footage.

Needing to "caption" the images in order to control how viewers received them, personalities on both networks frequently expressed their disgust. On CNN, Paula Zahn declared that the video had "made [her] sick," Aaron Brown called it "vivid and horrible," Carol Costello deemed the images "disturbing," and Lou Dobbs described the video as "gruesome" (CNN March 23 and 24). Fox News Channel personnel echoed those sentiments, calling the tape "disturb-ing," "chilling," "grotesque and sick," "utterly, utterly gruesome," "revolting and sickening," "monstrously gruesome," "garish," "quite horrific," "shocking," and "deplorable" (Fox News Channel March 23 and 24).

Despite the fact that state-run Iraqi TV was responsible for the production of the tape, CNN and Fox News Channel focused their attention on Al Jazeera for broadcasting it. Aaron Brown proclaimed the tape to be "far more horrible than anything we would ever show . . . on American television" (CNN March 23). Congressman Jim Gibbons, appearing on Fox News Channel, stated that he was "disgusted with Al Jazeera" because they were "breaking every rule" (Fox News Channel March 24). No one ever explained the "rules" of American television about this sort of traumatic event; broadcasters simply assumed that all U.S. viewers shared their sense of where the boundaries of propriety lay concerning images of U.S. soldiers. Of particular importance in this incident was the Pentagon's request that Al Jazeera stop airing the footage, ostensibly so military officers could notify the families of the dead and captured soldiers. With this action, the Pentagon stepped in to halt the transmission of images by a news organization whose audience was not based in the United States.

Surely the Pentagon and all of the networks involved in airing and exco-riating those who had produced the images surmised, as Giroux puts it, that in the context of "a media saturated with images of American violence against Muslims," the POW video was the beginning of "a means of response and a measure of revenge in the Arab world" (2006a, 49–50). POWs are a fact of war; U.S. networks had shown their own abundant footage of Iraqi POWs, some with guns pointed at their heads. But the fact that the enemy of the United States produced a video of dead and captured U.S. soldiers and submitted it for circulation on satellite news networks was a fundamental challenge to the staid and sanitized images supplied by U.S. networks. If the video produced by the murderers was an act of "revenge," then Al Jazeera's decision to air it was an act of balance (Fahmy and Johnson 2007).

It was also a matter of audience. In 2004, Shahira Fahmy and Thomas J. Johnson conducted a Web-based Arab-language survey to determine how viewers of Al Jazeera living in fifty-three countries[5] perceived the news outlet's signature graphic imagery. Eighty-seven percent of the respondents felt that Al Jazeera should show graphic images, and 81.4 percent felt that viewing graphic images of the Iraq war was a "good decision" (ibid., 255–256). One respondent did not consider the images to be "unpleasant"; rather, this individual likened the images to the "naked truth" that journalists should always unearth (256).

In hindsight, the POW video marks an intriguing precursor to the photographs taken at Abu Ghraib. Both sides were obviously capable of the same type of atrocities. Even though one set of images was meant for mass distribution and the other set was clearly intended for personal gratification, both groups chose to document their crimes. The pictures that emerged contested the veracity of the mainstream media's version of the war.

## Conclusion

During the 2003 invasion of Iraq, the spectacular look and sound of technology dominated war coverage. As Guy Debord maintains, the images and sounds themselves were not the spectacle. The power of the constructed look and sound to contain discourse and deny the material consequences of the war made them a spectacle, which Debord defines as "the existing order's uninterrupted discourse about itself" (1967, 24). Rarely did the war coverage challenge the existing order's dominance; the spectacle remained intact in the discourse of the news.

The graphics, animation, and embedded reports the two networks broadcast used the latest technology to put the superiority and hardware of the U.S. war apparatus on display. CNN and Fox News Channel journalists praised

---

5. While they lived in fifty-three different countries, 98 percent of the 638 respondents were originally from twenty Arab countries plus Afghanistan and Pakistan.

the technological might of the United States while they derided the outdated machinery of the Iraqi military. The visual celebration of weaponry bolstered the part of the war narrative that assured viewers that the United States could achieve its goals. Animated sequences were constructed with the assumption that the Iraqis would attack the United States with banned weapons and that the United States was ready with much more sophisticated hardware. The animation promoted the notion that the Hussein regime was a threat and the Iraqi military was a foe, if not a completely worthy one. The sequences answered the question "How can the Iraqis be a formidable enemy if we have such a magnificent arsenal?" by claiming that the Iraqi military had dangerous weapons too—illegal Scuds and hidden chemical weapons. Those claims sustained the momentum of the war narrative.

Technologically advanced ways of seeing, hearing, showing, and telling coalesced in a war narrative that tried to bring the interests of news viewers in line with the interests of the U.S. military. Up to March 24, Fox News Channel had exhibited a high level of adherence to the war narrative that CNN had not demonstrated. CNN never wavered from the narrative in any significant way, but its studio news commentators seldom expressed the brash enthusiasm of the Fox News Channel personalities. The networks' different approaches to their respective visuals and sounds provided a balance in their participation in the war narrative, however. Fox News Channel employed an aggressive verbal discourse, but CNN depended on its visuals to do the talking.

Driven by distinct visions, both networks maintained the sleek high-tech look and sound of high concept. Their allegiance to catchy theme music, title sequences, and graphics that promoted the ideology of the official war narrative completed the style of high concept. Frequently, sounds and imagery were used to excess, pushing the tools of the U.S. military to the fore and pulling back the narrative. Though complicated by narrative inconsistencies about the nature of the Iraqi threat, the elements of high-concept style explored in this chapter never abandoned the marketable concept.

# 6

## *The Marketing of the 2003 Invasion of Iraq*

In 2003, toy manufacturer Dragon Models created a series of action figures modeled on the U.S. soldiers in Iraq. One of those figures was named "Cody," and his specific assignment was special operations in southern Iraq (Dragon-ModelsLtd.com 2003). Although Cody was available commercially, Dragon Models crafted him for the Army and Air Force Exchange Service, which sells discounted products to military personnel and their families and donates the proceeds to the Army and Air Force (Taranto 2005). According to a Dragon Models marketing coordinator, Cody was "meant to look like a U.S. soldier who might be serving in Iraq." That likeness proved to be convincing. In 2005, the militant group Al-Mujahideen Brigade claimed to be holding a U.S. soldier captive and posted a photo of the hostage on the Internet as evidence. The photo appeared legitimate to the untrained eye, but upon closer inspection, military officials immediately noticed the so-called hostage's nonregulation military attire and detected the hoax. Dragon Models ultimately confirmed that the "soldier" was indeed Cody, whose toy assault rifle was pointed at his own head in the photo (CNN.com 2005). As an action figure, Cody represents the toy industry's take on the war. As a convincing decoy, he illustrates the scope of high-concept merchandising.

This chapter examines the 2003 invasion of Iraq as a commercial venture embraced by CNN, Fox News Channel, and other enterprises that exploited coverage of the war. CNN and Fox News Channel marketed their coverage and, in many ways, the Bush administration's premises for the war using promotional strategies during news programs and commercial breaks. Tie-ins from high-ticket merchandise like weaponry to more common consumer goods like action figures played a large role in selling the 2003 invasion to a U.S. audience. Of the 1991 Persian Gulf War, Jean Baudrillard writes, "The media promote the war, the war promotes the media, and advertising competes with the war. Promotion . . . allows us to turn the world and the violence of the world into a consumable substance" (1995, 31). CNN and Fox News Channel's coverage of the 2003 invasion of Iraq was one such "consumable substance" that in turn helped to commodify particular elements and experiences of the war.

## High-Concept News versus High-Concept Films

Like the turn toward high-concept filmmaking in the 1980s, the presence of high-concept values in news programming is a result of media conglomeration. The first wave of conglomeration in the 1960s aimed to decrease financial risk and increase the likelihood of substantial hits. Since the second wave in the 1980s, media conglomeration has been the active site of synergy, whereby the conglomerate promotes one media product across many of its holdings. Synergy is one way to achieve the maximum exposure a film requires to peak and sustain viewers' interest. Synergy, therefore, enables film companies to market movies aggressively and fuse a movie's style and content with its promotional life almost seamlessly. Other developments in the fields of marketing, technology, and film distribution enabled high concept films to thrive, and those same types of developments—combined with conglomeration—have likewise fostered the existence of high-concept news.

High-concept news is a more comprehensive manifestation of entertainment-driven journalism. In chapter 1, I discussed "infotainment"—a type of news that Thussu characterizes as "high-tech" with "complex graphics" and a style similar to video games (2003, 117). While infotainment accurately describes a contemporary trend in television news, the term high concept properly contextualizes and expands our understanding of the commercial reach of news.

What distinguishes high-concept advertising campaigns from all others is a marketable concept that is tailor-made for cross-fertilization in all the components of a corporate conglomerate (Wyatt 1994, 106). Advertising is the platform on which the excessive style of high-concept films makes sense and comes together. High concept is best understood as both an efficient packaging of a media artifact for maximum commercial gain and a strategy to keep the artifact in circulation for as long as possible, incessantly promoting itself and the system that allows it to circulate. Marketing and merchandising attempt to achieve those ends in high-concept news. During the first days of prolonged concentrated news coverage, as in times of war, viewership is high and television news outlets promote their coverage in ways similar to that of high-concept films.

However, the relationship between coverage of the 2003 war and high-concept marketing and merchandising is complicated by the simple fact that news coverage of a war is not a feature film. Obviously, some aspects of promoting a film do not translate wholesale to the arena of television news. One major difference is that the marketing of the 2003 war coverage involved an overtly political sell. The fixation of the networks on the procedures and tools of war—and not their deadly effects on human beings—infused their coverage with a celebratory anticipation of the unfolding events. Additionally, the majority of merchandising did not originate with the networks. That is not to say that the war coverage was devoid of commercial promotion. Indeed, CNN and Fox News Channel made a genuine effort to market the war, just as high-concept films promote consumer lifestyles.

The marketing and merchandising of the war depended on the creation of the ideal marketable concept, or the "hook," in Wyatt's terms (1994, 20). That hook grew out of the rhetoric of the Bush administration, which argued that there were undeniable connections between the September 11, 2001 attacks, Al-Qaeda, the Hussein regime in Iraq, and WMDs. The invasion of Iraq would produce a recognizable enemy (to replace the elusive Osama bin Laden), assert U.S. dominance to correct the nation's vulnerability on September 11, and bring democracy to the Iraqis. The U.S.-led coalition would execute a "humane effort" using high-tech weaponry that would spare the lives of Iraqi civilians and usher in a new era of democratic freedoms in Iraq (U.S. Department of Defense 2003d). Created by the Bush administration and the Department of Defense and maintained by CNN and Fox News Channel, the marketable concept of the 2003 invasion was the execution of vengeance through technological and moral superiority. CNN and Fox News Channel used different promotional techniques to communicate the marketable concept and to unite it with the networks' visions for their war coverage.

## Self-Promotion

Within the coverage of the war, promotional advertisements at CNN and Fox News Channel communicated an "image" in much the same way that ad campaigns for high-concept films do. In the film world the image conveys what the concept is, what genre it belongs to, and who stars in it (Wyatt 1994, 133). Media scholars argue that news networks have adopted many of the tactics of film companies to promote their products. Robert Stam writes that news programming "imitates the commercials by advertising itself . . . and by exploiting some of the same manipulative procedures" and techniques as commercials (1983, 36). Richard Cohen sees a link between the development of a feverish editing pace in commercials and the adoption of those techniques in news reports. The resulting content, he argues, is "practically subliminal" (1997, 37). Programming merges with the commercial aesthetic during commercial breaks in "boundaryless flow," a technique that McChesney attributes to the need for news programs "to satisfy those paying the bills" (1999, 51). News programs and commercials flow into one another so that viewers will resist the urge to channel-surf during breaks or avoid advertisements altogether. The flow of television programming combines programming, promotional spots, and commercials in a fluid sequence rather than offering hour upon hour of disjointed units (Williams 1974, 91).

Nick Browne argues that a complete understanding of and engagement with a television text—a news program, for instance—requires an awareness of a supertext—the relationship between the program, the industry, and the marketplace (1994, 71–72). If we look at coverage of the war at CNN and at Fox News Channel as competing supertexts, we can begin to see the intimate relationship between the units of "news" and "commercials," which seem to be discrete entities but actually are not. One example of the supertext is a network's use of promotional spots. The promo is placed at either end of the commercial break so that it will not interrupt commercials. It acts as a transition that takes on the style of the ads. The promo either weans the viewer away from news coverage to move into the more explicit commercial terrain or it reacquaints the viewers with the coverage they left just minutes before. In both cases, the news interacts with commercial breaks to achieve a commercial look in a way that is similar to the convergence of the high-concept film with its marketing campaign. By adopting the style of advertisements, news programming becomes a greater part of the commercial environment than it already is. During the coverage of the war, news networks inserted action sequences of U.S. soldiers in Iraq into their promos, removing the military campaign from its broader geopolitical context and engulfing it in the structure of a short commercial break.

After an initial period of commercial-free coverage, CNN and Fox News Channel resumed their commercial breaks on March 22. The number of breaks varied according to the pace of the day's breaking news, and the overall number of commercials was lower than usual across the cable news landscape. CNN aired commercials at a level that was only 50 percent of its normal rate, while MSNBC operated at 70 percent of its prewar levels (Greppi 2003). At Fox News Channel, most breaks included the networks' promotional spots, blurring the difference between news programming and blatant commerce. These promos sustained the marketable concept outside the context of traditional news coverage. In effect, the techniques that CNN and Fox News Channel news personnel used to promote the events and machinery of the war constituted another form of commodification. The world of war that CNN and Fox News Channel projected was in many ways the world of the commodity.

### Promotional Spots

During the initial days of the 2003 invasion of Iraq, CNN and Fox News Channel aired promotional spots that emphasized the work of their embedded journalists (see appendix C). Fox News Channel's promo packed considerably more information into thirty seconds than CNN's did, but both networks employed many of the same motifs. Both promos conveyed the genre of coverage (war), the characters (soldiers), and some of the stars (Bush and embedded reporters). The use of maps of Iraq and the region in general alerted the audience to the location. The information in these promotional spots gave viewers enough information to piece together a rough narrative.

In CNN's spot, the voice-over was as follows:

> CNN puts you alongside the troops. With soldiers: Walter Rodgers, Dr. Sanjay Gupta, Ryan Chilcote. With sailors: Frank Buckley, Kyra Philips, Gary Strieker. With airmen: Gary Tuchman, Bob Franken. With marines: Jason Bellini, Art Harris, Martin Savidge. Plus, CNN joins forces with top news organizations and our network of affiliates to put more reporters in the region than anyone. No one gets you closer than CNN—the most trusted name in news. (CNN March 22)

The promo featured CNN embedded reporters in various shots that put them close to the action of war. Martin Savidge sat alongside a military vehicle and a soldier; Walter Rodgers reported next to several military vehicles; and Dr. Sanjay Gupta, wearing a flak jacket, stood in front of stacks of sandbags.

Fox News Channel aired a comparable 30-second promotional spot during March 2003 with the following voice-over:

> Fox News Channel. The country at war. Stay with us for breaking news and live updates, fair and balanced, exclusively from the team you trust: Fox News Channel. On the ground. In the air. Reports from the front. Inside the conflict.

War coverage, second to none. Fox News Channel. The political fallout. With eyes around the world, a commitment here at home. The first place to turn for the latest in news—Fox News Channel. Real Journalism. Fair and balanced. (Fox News Channel March 24)

Both promos placed viewers in the position of the U.S. soldiers by providing them with the soldiers' points of view. The CNN promo showed two soldiers pushing an inflatable raft onto a ramp, shots from the point of view of a jet's cockpit, and soldiers walking into an abandoned building. The Fox News Channel promo also visually privileged the military's point of view. In it, viewers saw a full-frame shot from a tank's point of view that placed them behind the tank's gun; a shot of a soldier behind a tank's gun; a canted shot of a tank rolling from the bottom left of the frame to the top right; a shot from an airborne pilot's point of view; and a shot of soldiers' legs running in the desert, followed by a wider shot of those soldiers engaging in military exercises.

The images, which at times passed too quickly to be discernable without slow motion, were bold and active. Even if the images were of ordinary events —Fox News Channel's Mike Tobin walking through an Arab market, for instance—the pace of the editing determined the level of excitement. In less than two seconds during Fox News Channel's promo, viewers saw a close-up of a soldier, a shot of another soldier looking intently into a tank periscope, an exterior shot of a tank firing, and a shot of a jet flying away from an aircraft carrier. The pace of the editing heightened the drama of coverage, the involvement of the embedded journalists, and the mission of the military.

The final major sequence in Fox News Channel's promo involved the political aspects of the war, with shots of President Bush at a podium, Ari Fleischer at a podium, and a very low-angle shot of General Myers at a podium. The architects and mouthpieces of the war effort appeared quickly and in positions of power within the frame. No protesters or dissenting politicians appeared. The Bush administration's dominance in Fox News Channel's coverage of the war was apparent.

CNN and Fox News Channel also promoted products of their parent companies. CNN's promo, for example, advertised publications owned by Time Warner as well as other media outlets with which CNN had news-sharing agreements. When the promo's voice-over finished recounting the names of embedded journalists, a shot of a reporter in front of a helicopter with the U.S. flag waving in the background appeared. The on-screen text was the *TIME Magazine* logo. The next image revealed a group of soldiers peeking out of the top of a tank; the accompanying text was the *New York Times* logo. The logo of the *Boston Globe* appeared alongside an image of a soldier opening the cockpit door of a plane. Finally, the logo of *The Atlanta Journal-Constitution* appeared in conjunction with a soft close-up of a soldier that gradually came into focus.

Fox News Channel's promo, too, involved a significant commercial element. At the end of the promo, a camera in the desert focused on a commercially available Hummer (not a military Humvee) driving toward it. As the Hummer passed by the camera, the video slowed so that the civilian driver (an embedded reporter) became visible. The text "The Latest News" appeared as the Hummer passed by the camera. The Fox News Channel logo appeared, this time with another logo frame that included the tagline: "Real Journalism. Fair and Balanced." This type of product placement merged promotion of the war and promotion of a commercial product in footage that prepared the audience for its entrance into the standard commercial break.

CNN's and Fox News Channel's promotional spots devoted as much time to promoting the military effort and, by extension, the war effort, as they did to promoting the networks. The shots and the way they were arranged brought the actions and weapons of the military and the coverage of embedded journalists to the fore. The final shot of CNN's promo featured (possibly staged) footage of a soldier running with his back to the camera. A cameraman with his back to another camera pursued the soldier. The shot contained the kind of tension and excitement that made for a compelling visual, but it also performed a narrative function. Prior to this shot, the promo had established the location, the genre, and the characters and had established some stars and a rough narrative. The action shot at the end reminded viewers that in this war, reporters played a more central role in the narrative and in promoting that narrative than in previous wars. The commercial value of embedded reporters was obvious in these promos. As stars in this drama, they needed to be visible throughout the coverage and even afterward.

### Built-in Promotional Spots

CNN and Fox News Channel aired a number of other types of built-in promos throughout the initial days of the 2003 invasion. CNN boasted a detailed Web site devoted specifically to the invasion of Iraq titled "Special Reports: War in Iraq." CNN used its text crawl at the bottom of the television screen to promote the branded Web site, which carried links to other sites owned by Time Warner such as CNN International, Headline News, and CNN TV. The text crawl invited viewers to visit the Web site: "Go behind the scenes with CNN correspondents on the front lines at CNN.com"; "Go to CNN.com for 3-D models of stealth fighter, tomahawk missile, and other weapons in U.S. arsenal"; "What weapons of mass destruction is Iraq suspected of having? An interactive break-down at CNN.com"; "Go to CNN.com's war tracker for a daily briefing on the war in Iraq, with the latest info, maps, statistics & other news"; and "What areas have coalition troops secured in Iraq? Find out with CNN.com's war tracker" (CNN March 20–23). These text crawls promoted CNN's branded Web site, the war effort, and the war narrative. They identified Iraq in terms of the WMDs it supposedly was harboring and identified the United States in terms of its superior weaponry and pre-written conquest.

Like every other news network and, indeed, like fictional programs, CNN and Fox News Channel used teaser devices in their coverage to hold the attention of their audience. Stam discusses the ways in which the news "titillates our curiosity and keeps us tuned in and turned on," but Fox News Channel's built-in promos exceeded the usual "coming up" moments on television news (1983, 32). Shepard Smith, Mike Emanuel, and E. D. Hill transposed teasers onto the developing war events. Before the "shock and awe" bombing campaign, both networks covered the anticipation of the event and the frustration at its delay. On March 20, Smith commented, "What we have not yet seen and what is promised is a campaign of shock and awe. . . . There'll be no mistaking it, and we'll be here for you to cover it" (Fox News Channel March 20). Smith's comment built up the campaign and used the bombing as a device to promote Fox News Channel's coverage. Emanuel did likewise when he assured viewers that Fox News Channel's "cameras [were] all pointed in the right direction" (Fox News Channel March 21).

On two occasions on March 24, Hill used an imminent bombing as a built-in promo. The first time Hill used this tactic, she suggested that viewers take note of the time that B-52s were taking off from an airbase in the United Kingdom and then add six and a half hours (the time it would take the aircraft to reach Iraq). Hill said, "You just make sure that you're tuned to the TV set because that is when *live* you will see what happens in Baghdad" [her emphasis]. Three hours later she inserted another teaser: "In just about three hours you want to make sure you are tuned to Fox News Channel because at that time, if they're going after Baghdad, the B-52s should be right about there" (Fox News Channel March 24). This differed from a typical teaser, as it did not promote something relatively benign like an upcoming edition of *The O'Reilly Factor*. Fox News Channel journalists explicitly used a destructive bombing run—"going after [a city of 6 million]"—as a promotional tool, an approach that seemed to contradict Hill's statement on the same episode of *Fox & Friends* that "[Fox News Channel news personnel are] not a pro-war or bloodthirsty bunch" (Fox News Channel March 24).

Other self-promotion came in the form of guests, whose presence synergistically promoted other brands within the networks' respective conglomerates. For instance, CNN studio personnel frequently carried on phone conversations with the embedded reporters of its corporate sibling *TIME Magazine.* When that happened, the *TIME* logo always appeared on screen. Bill Kristol (who edits the *Weekly Standard,* owned by Fox's corporate parent News Corporation) and guests from Sky News (News Corporation's British news channel), were similar fixtures on Fox News Channel. These incidents created multiple opportunities for cross-media promotion and demonstrated how the parent companies of the two networks insert their corporate interest in the content of the news.

## The Benefits of Self-Promotion

CNN and Fox News Channel engaged in self-promotion during commercial breaks and news programs and through merchandise connected to their coverage of the war. In July 2003, CNN released the video *CNN Presents: War in Iraq—the Road to Baghdad.* One of the many items Fox News Channel advertised was the video *War Stories from Iraq with Oliver North,* which is part of a series of videos about military history. This particular volume dealt specifically with the invasion of Iraq in 2003. A book entitled *War Stories: Operation Iraqi Freedom* accompanied the video. The promotional materials referred back to the network's coverage of the war and to the narrative it constructed, solidly linking the network's news to commercial space.

The most significant benefit of the promotion came in the form of ratings, however. During March and April 2003—the first two months of the invasion of Iraq—CNN and Fox News Channel had their largest audiences since September 2001. In March 2003, CNN surpassed its September 2001 viewership of 2.1 million viewers, drawing 2.3 million viewers—a number that fell to 2.2 million the following month before dropping to below the 1 million mark again in May. Fox News Channel's audience size surpassed CNN's by almost 1 million viewers in March. The 3.2 million viewers that chose Fox News Channel in March rose to 3.5 million in April and fell to 1.6 million in May (Project for Excellence in Journalism 2005, "Audience" section).

## Promotion of the Military and the War

No clear distinction exists between the two networks' promotion of their coverage and their promotion of the war. I distinguish between self-promotion and war promotion based on degrees of emphasis. In both types of promotion, CNN and Fox News Channel maintained the marketable concept and their respective styles, but they promoted different products for different beneficiaries. Former ABC correspondent Beth Nissen called the 1991 Persian Gulf War coverage "an advertisement for the U.S. military: the bombs always hit the targets, the U.S. government always scored perfectly. The pictures were like Pentagon commercials and we just played them" (Andersen 1995, 212). If we consider Nissen's claim in light of promotion of the 2003 war at CNN and Fox News, we can conclude that the biggest beneficiaries of the networks' promotion of the war were the official proponents of the marketable concept.

*Code Names*

The most obvious instance of war promotion came in Fox News Channel's use of the military's code name for the invasion, or what one might consider the brand name of the war. T. G. Riese & Associates, a consulting firm that helps companies develop strong brands for their products, noted that "every war needs a compelling brand proposition, whether or not it is publicly acknowledged" (Tomkins 2003). The goal of war branding is the same as the goal of all branding—to create and maintain a strong relationship between consumers and products. The most visible manifestation of war branding is the use of code names for military operations, a practice that began in covert operations after World War II (Blumenstein and Rose 2003). Code names were available for media consumption in 1989 when the United States invaded Panama (Laing 2003). Since Operation Just Cause in Panama, the U.S. has used such code names as Desert Shield, Desert Storm, Restore Hope, Uphold Democracy, Shining Hope, Determined Force, Provide Promise, and Enduring Freedom (Laing 2003). Carolyn Said (2003) writes that the 1991 Persian Gulf War "is widely considered the first instance of a brand-centric war, consciously tailored for the mass media." She also argues that the military and the U.S. government used the devices of advertising much more extensively in 2003 than they had in the 1991 war. A clear example of that practice was White House Chief of Staff Andrew Card's statement about why the invasion began in the spring: "From a marketing point of view, you don't introduce new products in August" (quoted in Johnson 2003).

An even clearer example, and one that leans toward high concept, was the code name given to the operation that eventually located Saddam Hussein: "Red Dawn." The operation was intentionally named after *Red Dawn* (1984), a movie about a group of teenagers who fend off a Soviet invasion of their hometown in Colorado. The code name even prompted a trivia question on CNN that asked, "Which military operation is named after a Hollywood movie?" (CNN December 16, 2003). News organizations do not have to incorporate official operation code names into the titles they assign to their war coverage. Nevertheless, Fox News Channel wholeheartedly accepted the official code name that communicated part of the marketable concept—"Iraqi Freedom" —and packaged its 2003 war coverage accordingly. Fox News Channel's adoption of the code name as its war coverage title was transparently promotional.

## Soldiers (Retired and Active)

Another conduit for war promotion was military personnel, who typically "spare journalists the trouble of looking for people who really have something to say" (Bourdieu 1998, 30). In their coverage of the 2003 war, CNN and Fox News Channel relied heavily on military analysts and thus privileged the military's point of view. Fox News Channel had at least nineteen military analysts on the payroll, according to its Web site's list of contributors, and CNN had approximately five permanent military analysts. The constant rotation of in-house military analysts at both networks kept commentary squarely focused on military strategy rather than on the effects and the social or political context of the war.

At Fox (but not at CNN), network advertisements also promoted the war effort. Two promotional campaigns merged the identity of the network with the U.S. soldiers fighting the war. Beginning on March 21, Fox aired a series of spots entitled "Heart of War." Jon Scott introduced the first "Heart of War" spot with this statement:

> It is said that nobody hates war more than the people who have had to fight in them, but if you spend any time around our professional soldiers ... you know that they are very well-trained, very smart, and very patriotic. We wanted to give you a chance to meet some of these folks. Take a look now at the heart of war. (Fox News Channel March 21)

Each spot featured a soldier wishing his family and friends well and concluded with the soldier making a plug for Fox News Channel.

An additional set of promotional spots was reminiscent of commercials for companies like Qwest, whose "the spirit of service, there for you" campaign featured and praised its reservist employees (Ellin 2003). This set of Fox promos featured slow, respectful music and still photography. In lieu of voice-over, on-screen text read: "To our armed forces. For your courage, for your sacrifice, for your bravery, we salute you" (Fox News Channel March 22). The first photograph featured a portion of a soldier's dress uniform in the light while the rest of the frame was shrouded in darkness. Next was a photo of soldiers hanging from a helicopter. After that was a photo of a pilot looking out a window. Similar photos followed; some were action shots and others were more introspective. The promo concluded with the Fox News Channel logo. Fox News Channel carried at least three versions of this type of promo, aligning the network undeniably with the troops and reinforcing the soldiers' heroism and the network's patriotism.

*Weapons (Part I)*

Both networks contributed to the marketable concept by promoting the weapons in the U.S. arsenal. CNN posted an extensive section on U.S. and Iraqi weapons on the special war coverage page of its Web site entitled "Special Reports: War in Iraq," which devotes a great deal of space to military matters. The main menu of the site includes the categories War Tracker, Forces (which included the three subcategories of U.S. & Coalition, Iraq, and Weapons), Maps, On the Scene, Sights & Sounds, Impact, Heroes of War, and Struggle for Iraq. In the section entitled Forces: U.S. & Coalition, the Web site chronicles the units involved in each branch of the U.S. military; lists the names of the U.S., British, and Australian commanders; lists the members of the coalition forces; and provides a full list of U.S. weapons. The site provides comparable information for the Iraqi military.

By far the most detailed part of the Web site is the section devoted to coalition and Iraqi weapons. The site breaks the weapons down into munitions, ground weapons, aircraft, warships, and WMDs. Three-dimensional models of select U.S. munitions, aircraft, and ground weapons are available as well. For instance, clicking on the hyperlink for the Massive Ordnance Air Blast bomb opens another window with a picture of the bomb and its relevant statistics. Clicking on the hyperlink for an aircraft carrier such as the USS *Constellation* reveals an entire page of statistics, images of the carrier battle group, and a large photo of the *Constellation*. The sophistication of CNN's weapons site approaches the sophistication of on-air graphic representations of weapons and aircraft on both networks (see chapter 5 and appendix B).

The networks' promotion of the marketable concept required them to promote the U.S. military and the general war effort. CNN and Fox News Channel performed these promotional tasks differently but with similar energy. Fox News Channel adopted the official code name of the operation. Fox News Channel's pro-troops promotional spots and CNN's military-heavy Web site marked an interesting divide. In spite of the fact that both networks emphasized the marketable concept, Fox News stressed an emotional connection to the soldiers in its promos, while CNN's Web site focused on the technology the soldiers used. CNN's display of weapons functioned to promote the marketable concept in two ways: it celebrated high technology, glossing over the lethal consequences of weaponry in the process, and it served as a form of merchandising.

## War Sells

Merchandising functions most successfully in the service of films that are based on bold images, hence the hyperstylization of high-concept films. The field of film merchandising includes both products officially licensed to promote the film and products that function as ancillary tie-ins. The key distinction is that ancillary tie-ins do not benefit the studios directly. They are unlicensed, so their value to the studio stems from their ability to create and sustain awareness of the film. To illustrate the value of tie-ins, Wyatt discusses the fashion trends inspired by *Flashdance* and *Saturday Night Fever* (1994, 153). "Inspired-by" looks are not connected to the films' producers legally, but they promote the films nonetheless. The applicability of ancillary tie-ins to the coverage of the war in 2003 is substantial. The merchandising that circulated outside the networks—including the marketing of products such as weapons, automobiles, action figures, and video games—supported and sustained the war narrative and the marketable concept. These merchandising exploits were made possible in part by the war narratives and the coverage of the war at both networks. The ideas that those networks helped to create will remain in circulation because of these products.

### Weapons (Part 2)

The ways that CNN and Fox News promoted weapons assume another level of significance when one considers the international market for those weapons. A 2003 *Washington Post* article discusses how the war functioned as an advertisement for the products of U.S. weapons producers in the international market. The coverage of the war highlighted the fact that these munitions were now "battlefield-tested weapons." The director of arms sales monitoring at the Federation of American Scientists went so far as to say that CNN was "the best marketing tool" for the weapons trade (Merle 2003). That sentiment is hardly new. Andersen writes that in the coverage of the 1991 Persian Gulf War, weapons became like the products in television commercials that could perform feats of magic: "Weapons became the magical objects that, in the most simple, clean, and easy way, would solve the problem in the Persian Gulf" (1995, 217). Engelhardt argues that the 1991 war was a "forty-three-day-long ad, intent on selling both domestic and foreign markets in the renewal of U.S. qualities, as well as on the specific weapons systems that were in the process of renewing those qualities" (1994, 87). These trends in war coverage were replicated in 2003; Merle notes that the weapons-heavy coverage of the 2003 invasion reproduced the marketing environment that boosted arms sales after 1991 (Merle 2003).

The types of weapons in high demand were the satellite-guided precision weapons that CNN and Fox News Channel devoted so much time to describing (ibid.). While network personnel focused on how precision-guided munitions had the capacity to prevent massive civilian casualties, Merle reminds us how that type of coverage might benefit the arms trade. Indeed, sales of such munitions increased dramatically internationally in 2003 (Berrigan, Hartung, and Heffel 2005).

## Toys

Toys that targeted children and adults were a significant part of the merchandising of the 2003 invasion of Iraq. Military-themed action figures and video games replicated the narrative and character types represented on CNN and Fox News Channel. These toys contributed to the high-concept merchandising of the war coverage by latching onto the marketable concept the news networks promoted. One film executive asserts that a media text "must develop into toyetic applications—characters which have a personality that can be easily transferred to dolls and playset environments" (Wasko 1994, 207). The way that a high-concept film can spawn toys that transparently promote the film is a testament to one of the fundamental characteristics of high concept: ease. The easy pitch and the easy sale translate to an easily understood narrative and easily understood characters. The rhetoric from the White House and the networks seamlessly and easily promoted the high concept of the war, and the toys soon followed.

One of the biggest lines of merchandise to spring from high-concept films is the action figure. The earliest action figures were military-themed toys, which have evolved from basic wooden figurines to detailed plastic figures that are tailor-made to signify the latest military conflict. The influence works both ways: astoundingly, one Pentagon spokesman has even admitted that the M-16 and a number of other items in the military arsenal, including drones and assault weapons, were based on creations by toy manufacturer Mattel. A palpable connection exists between the military, toy manufacturers, and news networks, as an exchange at a toy fair early in 2003 indicates. An Army and Air Force Exchange Service buyer recommended to some toymakers that they wait until the invasion began and then use the same logos that would appear on CNN once the war started. That way, the toys would be more specific to the current conflict. As instructed, toymakers eagerly watched the news coverage for "new battle tools" that might prove profitable as "new battle toys" for the upcoming Christmas season. The presidents of toy manufacturers Small Blue Planet and Plan-B Toys even acknowledged that news coverage of the countdown to the invasion drove new toy production (Hamilton 2003).

Older toys still in circulation also received a boost from the buildup to the war. Sales of Hasbro's GI Joe line increased before and during the 2003 invasion. Noting that Hasbro had expected the increase, one retail spokesman remarked, "People get rather patriotic, and they go out and buy their GI Joes. It happened before the [1991] Persian Gulf War" (Associated Press 2003). Small Blue Planet's president, Anthony Allen, admitted that the company "started work when the 'Showdown' buzzword hit the airwaves"; the line of action figures that resulted was branded "Special Forces: Showdown with Iraq" (Hamilton 2003). Dragon Models named its series of Iraq-specific figures "Dragon Extra 4: Operation Iraqi Freedom." In addition to the Cody figure I mentioned at the beginning of the chapter, the line consisted of such action figures as "Fernando," an M-240B marine gunner on the highway to Baghdad, and "Walt," who performs reconnaissance on the outskirts of Basra.

Other action figures that were not soldiers also appeared on the Web sites of smaller toy manufacturers. For example, Blue Box Toys developed "Elite Force Aviator," also known as the President Bush action figure. The toy Bush appeared in the "traditional naval aviator garb" the president wore for his widely televised May 1, 2003, landing on the USS *Abraham Lincoln* (Jacobs 2003). And Herobuilders.com featured a variety of action figures of Bush, Iraqi information minister Mohammed Said al-Sahhaf, Saddam Hussein, and others. That Iraqi action figures existed at all was not a complete surprise; Karim H. Karim argues that the figure of the "violent Muslim" carries a lot of weight in popular culture. He writes, "Commercial and ideological purposes dovetail neatly in products that exploit the basic stereotypes of Arabs and Muslims" (2006, 120). That profitability was evident to toymakers. In an e-mail on January 10, 2006, a Herobuilders.com spokesman revealed to me that the information minister figure "sold the best in the shortest period of time" and that the Bush action figure had the second highest sales figures. The 2003 war coverage, rooted in formulaic narrative, genres, stars, and character types, proved to be very "toyetic."

Video games surpass even action figures in popularity. The video-game software industry claimed $5.8 billion in overall sales in 2003 (NPD Group 2004). Military-themed video games are a subset of strategy and first- and third-person shooter games. One of the titles at Gotham Games is *Conflict:*

*Desert Storm,* Sony Computer Entertainment has *SOCOM* (the acronym for United States Special Operations Command), and the U.S. Army promoted two video games on its Web site: *America's Army* ("the official U.S. Army game") and *Medal of Honor* (created by Electronic Arts). At the end of the first quarter of 2003, *SOCOM: U.S. Navy Seals* (released August 2002) had the eighth-best sales figures among U.S. video games (The NPD Group 2003a). *SOCOM II: Navy Seals* (released November 2003) debuted at number five, although it fell to ninth one month later (NPD Group 2003b, 2003c). *Medal of Honor: Rising Sun,* a World War II–themed game (also released in November 2003) debuted at number two (NPD Group 2003b, 2003c). Although it fell to number seven one month later, it was the tenth best-selling console video game of 2003 (The NPD Group 2004). In general, games set in the Middle East with "themes of terrorism and war" performed well in 2003 (WorldWatch Institute 2003).

The video game most relevant to the 2003 invasion of Iraq is entitled *Kuma\War* from Kuma\Reality Games. *Kuma\War* is a free downloadable game that offers print news stories, Department of Defense briefings, video footage of soldiers, and official information about operations conducted in Iraq in a first- and third-person shooter video-game format. The director of *Kuma\War*'s video services is William B. Davis, a former ABC News producer (Gaudiosi 2003). The game's video footage comes from Department of Defense photographers, the Associated Press and other wire services, broadcast television networks, overseas television stations, and "home video." Videos from the Department of Defense and the wire services figure heavily in the game because they are the least expensive to obtain (Anderson 2003).

The news reports in the video game function as bits of intelligence that players use to educate themselves about the missions they are about to play. The game's Web site offers this advice:

> Look at the Mission Intel Pack before you play and you'll learn what the 101st Airborne was up against when they captured the Hussein brothers or the 10th Mountain Division when they took on Al-Qaeda—then you can jump in the game and see for yourself—in single player or multiplayer, co-op or red on blue intense action. (Kuma\War 2004)

In the game, which boasts "Real War News," an actor portraying a news anchor presents the footage of soldiers as actual news. The news report acts as an introduction for each mission. Of eighty-three missions currently available, there are ten different missions set in Fallujah, one mission that portrays Operation Red Dawn, two missions that link Al-Qaeda and Iraq, and two missions devoted to the campaigns that resulted in the deaths of Uday and Qusay Hussein (Kuma\War 2008).[1]

Because it is not affiliated with CNN or Fox News Channel, *Kuma\War* demonstrated (and continues to demonstrate) the reach of the war narrative the two networks constructed beyond the conventional news arena. That the Department of Defense approved the use of footage for entertainment purposes in *Kuma\War* speaks to its willingness to perpetuate the official war narrative on a multiplicity of platforms. Furthermore, the participation of major news networks like ABC and former news personnel underscores the porous nature of the barrier between news and entertainment. The combination of actual footage, a "fake" news format, and a gaming context illustrates precisely to what degree the 2003 invasion has come to be constructed as a high-concept commodity. The narrative construction of the invasion by television news producers provided the formula such merchandise needed in order to thrive.

## Music

The marketing potential of recorded music was not tapped into on a large scale until the 1970s, though the film and music publishing industries began a lucrative relationship once sound technology put songs on film (Wyatt 1994, 134). The soundtrack market is now a given in the overall merchandising scheme of high-concept films, even when some soundtracks feature songs that were inspired by the films but were not actually in the films. The same pattern exists in times of crisis. As in the case of the September 11 terrorist attacks, the 2003 invasion of Iraq inspired a wealth of musical responses, both for and against the war. The pro-war songs are extremely faithful to the version of events posited by the Bush administration, CNN, and Fox News Channel. Each contains elements of the war narrative that the network war coverage emphasized, and two that were released before 2003 act as a transition between the war narrative's cause (the September 11 terrorist attacks) and its effect (military intervention). Four songs from the country music genre exemplify the value of the tie-in for the high-concept text.

One of the transitional songs, Toby Keith's "Courtesy of the Red, White, and Blue (The Angry American)," was the number-one single on Billboard's

---

1. Most of the missions relate directly to Iraq, but some—like Mission 73: Mexican Border Battle—focus on other themes related to national security (Kuma\War 2008).

"Hot Country Singles and Tracks" for two weeks in the summer of 2002 (Pietroluongo, Patel, and Jessen 2002). Although the song was released well before the invasion, it clearly links the September 11 attacks to the need for military action—a link that has persisted throughout the war in Iraq. At the beginning of the song, Keith sings, "Now this nation that I love / Has fallen under attack / A mighty sucker punch came flyin' in." He concludes with the threat of nationalistic vengeance:

> Justice will be served
> And the battle will rage
> This big dog will fight
> When you rattle his cage
> And you'll be sorry that you messed with
> The U.S. of A.
> 'Cause we'll put a boot in your ass
> It's the American way. (Dreamworks Nashville 2002)

The connection between September 11 and military retribution couched in jingoist terms persists in another transitional song, Charlie Daniels's arguably racist piece, "This Ain't No Rag, It's a Flag." He begins by alluding to a derogatory stereotype of Arabs: "This ain't no rag it's a flag / And we don't wear it on our heads." Like Keith, he concludes the song with the threat of retribution: "We're gonna hunt you down like a mad dog hound / Make you pay for the lives you stole" (BMG Heritage 2001). The Charlie Daniels song reached number thirty-nine on Billboard's "Hot Country Singles and Tracks" in 2001 (Pietroluongo, Patel, and Jessen 2001).

A third song that perpetuates the marketable concept as well as the war narrative's character types is "Have You Forgotten?" sung by Darryl Worley. The song was country music's number-one song for six weeks in spring 2003 (Pietroluongo, Patel, and Jessen 2003). Worley explicitly referred to September 11, 2001, at the beginning of his song: "Have you forgotten when those towers fell?" He concluded by linking those events to the war:

> I've been there with the soldiers
> Who've gone away to war,
> And you can bet that they remember
> Just what they're fighting for. (Dreamworks Nashville 2003)

Clint Black's "Iraq and I Roll" does not broach the subject of revenge, but its jingoistic lyrics invoke the technologically advanced U.S. arsenal and mention Hussein by name. In the middle of the song Black sings,

It might be a smart bomb
They find stupid people too
And if you stand with the likes of Saddam
One might just find you
I rock, I rack 'em up, and I roll
I'm back, and I'm a high-tech GI Joe
I've got infrared, I've got GPS, and I've got that good old-fashioned lead
There's no price too high for freedom
So be careful where you tread. (Clint Black 2003)

Military might and arrogant rhetoric joined forces in Black's song and all of the others that followed the Department of Defense's narrative as closely as CNN and Fox News Channel did.

These songs formed a soundtrack to the war as seen on television. The two songs released before the 2003 invasion served as a bridge between the September 11 terrorist attacks and the U.S. military action that followed in Afghanistan and in Iraq. The two songs released after the 2003 invasion repeated the same themes stressed on CNN and Fox News Channel, and they featured the same character types. They were the musical equivalent of some of the more extreme war coverage found on Fox News Channel; they brimmed with nationalistic pride and simplistically linked September 11, Al-Qaeda, and Saddam Hussein. Like the war coverage, the songs adhered to a formula that was eminently marketable, a necessary attribute for songs recorded on major labels.

Action figures, video games, and recorded music all performed similar functions during and after the 2003 invasion of Iraq. They were committed to the military stars—the soldiers. Some even went so far as to embody Iraqi stars. These ancillary tie-ins were valuable because they kept the motifs of the war coverage in circulation; Kuma\War actually circulated war coverage in the context of a computer game. In turn, uncomplicated and sanitized coverage of the war created interest in the products. The relationship was symbiotic, even if the producers on each side were not directly involved with the other.

## Conclusion

The specificities of television problematize a faithful adherence to the elements of high concept used in the film industry. The marketing and merchandising of war coverage on CNN and Fox News was understandably different from the marketing and merchandising of high-concept films like *Batman* or *Stars Wars*. However, one should never underestimate commercial television's basic need to market. The merchandising on the two news networks occurred on a small scale, but the networks' efforts to promote themselves and the war were considerable.

The merchandise that revolved around the war narrative ensured that it would saturate popular culture. The war tie-ins advertised the story, stars, character types, genres, and ideology of the high-concept war narrative. The success of the products provided evidence that the concept of the war was marketable.

# CONCLUSION

## *The Narrative Exits Screen Right, the Coverage Fizzles, and News Is What, Exactly?*

Kellner writes that as the events of the war began to get more chaotic—or as they refused to follow the Department of Defense's script—"the embedded and other on-site reporters provided documentation of the more raw and brutal aspects of war in telling accounts that often called into question official versions of events as well as military propaganda" (2005, 65). The narrative could not hold in spite of the best efforts of the Department of Defense and the networks. But discourse adjusted to meet the crisis. In 2007 and 2008, as calls for troop withdrawal intensified, the story became about Iraqis needing to step up and help themselves. Before the coherent narrative disintegrated on a massive scale, one media event—the climax of the narrative—underscored the growing rift between the way that CNN and Fox News Channel implemented the war narrative.

On April 9, 2003, CNN and Fox News Channel showed U.S. soldiers and tanks encircling Al-Firdos Square in Baghdad. In that part of the city, Iraqis appeared to walk the streets freely, coexisting peacefully with the soldiers who talked among themselves, patrolled the area, and eyed the rooftops of buildings suspiciously for snipers. CNN's Martin Savidge reported heavy fire in Baghdad, and throughout the morning, he transmitted coverage of gunfire at Baghdad University. Fox News Channel's Rick Leventhal also described sounds of gunfire a few miles away. Nevertheless, CNN's Paula Zahn and Fox News Channel's Jon Scott announced "spontaneous celebrations" in the square, while Zahn added that "total anarchy" reigned in other parts of Baghdad (CNN April 9; Fox

News Channel April 9).[1] The bulk of the coverage focused on Al-Firdos Square, however, and at approximately 10 AM ET, a series of events began to unfold that compelled Fox News Channel to drag the marketable concept through another major complication.

First, Iraqi men surrounded a statue of Saddam Hussein and pelted it with shoes. The U.S. Marines gave them rope to try to topple the statue, but the Iraqi men failed. The marines then used their M-88 Hercules tank recovery vehicle to secure a chain around the statue. After the chain was secured, a soldier passed a U.S. flag up to Corporal Edward Chin, who then covered the statue's head with it. The flag remained in place briefly, and then Chin replaced it with a pre-1991 Gulf War Iraqi flag, draping it through the chain around the statue's neck like a kerchief. That flag came down, too, and so did the statue.

Throughout the morning, CNN and Fox News Channel journalists watched and commented extensively, framing the event in terms of character types, liberation, and regime change. Before the flag incident, the gravity of the fact that U.S. troops were in Baghdad elicited claims of historical import. CNN's Christiane Amanpour and Bill Hemmer and Fox News Channel's Brigitte Quinn explicitly compared the event to the fall of the Berlin Wall. Fox News's David Asman, who claimed he had goose bumps, compared it to the tearing down of Lenin statues. Network personnel also commented on the energy in the square. Hemmer used the term "revolutionary" to describe the scene, but he said he would use the word "loosely." Paula Zahn exclaimed that the activity in the square was "really extraordinary to watch," and a corporal speaking on the phone with CNN observed "a lot of jubilation" by people who were "really happy" the United States was there.

Shots of the square did not permit an accurate head count of the jubilant onlookers. The corporal estimated a crowd of seventy. Later, a reporter transmitting to CNN increased that number to 1,500. Fox News Channel studio personnel claimed that "hundreds of joyful people" were present, yet ABC's Peter Jennings called the crowd "small" (Harper 2003). The crowd estimates varied, just as the reactions to the U.S. flag did.

The image of the U.S. flag on the statue's face did not coalesce with the liberation aspect of the war narrative, and the Pentagon and CNN made that

---

1. Unless otherwise indicated, all the quotes from transcripts of CNN and Fox News Network broadcasts in this chapter are from April 9, 2003.

clear. When First Lieutenant Tim McLaughlin passed the flag to Chin in Al-Firdos Square on April 9, Amanpour was noticeably exasperated. She asked, "What is he doing?" She continued, upset with the imagery, "Oh, there we have it, an American flag. Now this, I have to say, may look like fun, but a lot of the soldiers have been told not to sort of do this triumphalism because it's not about American flags but about Iraqi sovereignty and the Iraqi people, country, etcetera." Paula Zahn referred to the flag-raising at Umm Qasr, and she recalled that Iraqi civilians "were so angry about that show of what they called American imperialism. That flag came down within minutes." A short time later, Barbara Starr revealed that there had been "an almost audible gasp" at the Pentagon. Later in the evening, Wolf Blitzer stressed that images of occupation were not desirable, but in an interview conducted by Paula Zahn, an Iraqi-American citizen remarked that the flag on the face was "not a big deal." That reaction was apparently an aberration. Though viewers were not privy to shots of the crowd's reaction when the flag incident occurred, Starr commented that there was "not a good reaction from the people there."

Like the reaction of journalists at Fox News Network to the Umm Qasr flag-raising, the response to the Al-Firdos Square flag incident was euphoric. Chris Jumpelt exclaimed, "Here we go! The American flag! There we go! Saddam Hussein is now under the Star Spangled Banner. That's all you're gonna see from now on!" When Chin removed the flag, David Asman quickly attempted to rescue the situation:

> Look who's holding [the flag] now. It's an Iraqi citizen waving the U.S. flag. That might do something to dispel criticism no doubt we are about to hear from with regard to the Al Jazeera stations . . . . You can understand these Marines who have put their lives on the line, sweated with blood and guts for the past three weeks, wanting to show the Stars and Stripes in this moment of glory. Understandable, but no doubt Al Jazeera and the others are gonna make hay with that.

With that statement, Asman underscored the soldiers' heroism and the perception at Fox News that Al Jazeera's coverage of the war was biased. Asman also displaced the idea of occupation onto the Iraqi populace, which Fox News was ever ready to press into any interpretive frame necessary (even without evidence). In his view, the "Iraqis" (actually just one man) wanted the U.S. flag displayed just as they wanted the coalition soldiers in Baghdad. Earlier, Asman had declared the "heart of Baghdad . . . liberated," and for Fox news personnel, the participation of a single Iraqi in the flag incident made the liberation narrative complete in Al-Firdos Square.

Of the Hussein statue incident, Brit Hume opined,

> These pictures are the purest gold that the administration could imagine for this day in terms of putting to rest . . . the issue of how the Iraqi people feel about the mission and the war, and how they feel about Saddam Hussein. This speaks volumes and with power that no words can really match.

Hume pressed the visually present yet sonically absent Iraqis into the service of the network's war narrative, despite a glaring lack of evidence. Likewise, Brigitte Quinn asked, "Did the Iraqi people ever dream that they could do something like this?" Quinn used the palpable anger the group was directing at Hussein to justify the war narrative that said they welcomed U.S. "liberators," even though the first sentiment did not necessarily connect to the second.

At the Department of Defense briefing on April 9, which occurred well after the flag incident, Rumsfeld ignored the controversial display of the U.S. flag and focused instead on the success of the war narrative:

> The scenes of free Iraqis celebrating in the streets, riding American tanks, tearing down the statues of Saddam Hussein in the center of Baghdad are breathtaking. Watching them, one cannot help but think of the fall of the Berlin Wall and the collapse of the Iron Curtain. We are seeing history unfold events that will shape the course of a country, the fate of a people, and potentially the future of the region. Saddam Hussein is now taking his rightful place alongside Hitler, Stalin, Lenin, Ceausescu in the pantheon of failed, brutal dictators, and the Iraqi people are well on their way to freedom. (U.S. Department of Defense 2003g)

Fox News Channel personnel took their appraisal of the situation further than the Department of Defense did, but neither set of onlookers allowed the events to dismantle the narrative. In fact, their insistence that the Iraqis responded positively to the presence and actions of the U.S. military in Al-Firdos Square carried the marketable concept through an event full of contradiction.

For the two networks' on-air personalities, the war narrative was proceeding as planned, in spite of the continued fighting and images of actions of U.S. soldiers that contradicted the war narrative of the U.S. government. The actions of soldiers with U.S. flags at the former UN post, Umm Qasr, and Al-Firdos Square were thematically identical; they challenged the liberation aspect of the narrative and they illustrated the need to keep the narrative intact and safe from unsettling images. Yet the networks' different reactions to the flag incident at Al-Firdos Square—two weeks after the Umm Qasr incident—demonstrated the lengths to which Fox News Channel would go to uphold the marketable concept.

## Military Puppets

As it turns out, the Pentagon went considerably farther to uphold the marketable concept and the war narrative. Although it came as no great surprise, a
*New York Times* article in April 2008 revealed that the Pentagon, under orders
from the Bush administration, used military analysts as "a kind of media Trojan
horse" to feed news programs predetermined talking points about the invasion
(Barstow 2008). A former aide to Victoria Clarke said that the decision to use
analysts as public relations tools was made in the year *before* the invasion. The
Pentagon's roster included over seventy-five officers, most of whom worked for
Fox News Channel.

Using e-mails and memos as evidence, *New York Times* reporter David
Barstow found that the Pentagon had recruited approximately seventy-five retired military officers and used them to construct the administration's narrative
of the war as news analysts on network television. Many of the analysts were
lobbyists, senior executives, board members, and consultants for military contractors; these ties obviously compromised their objectivity. The Pentagon gave
this hand-picked group of military analysts access to classified information and
took them on tours of Iraq. Some of the analysts later expressed regret at the role
they had played in serving as government mouthpieces; Robert Bevelacqua, a
former Fox news analyst and a retired Green Beret, said "It was them saying, 'We
need to stick our hands up your back and move your mouth for you.'" Another
participant later acknowledged the lack of truth in the information he and his
colleagues had been fed; Kenneth Allard, who has taught information warfare
at the National Defense University, later said that what they had been told by
the Pentagon and what was actually true was "night and day," adding "I felt like
we'd been hosed" (Barstow 2008).

Barstow found evidence that "again and again . . . the administration has
enlisted analysts as a rapid reaction force to rebut what it viewed as critical news
coverage, some of it by the networks' own Pentagon correspondents." The Pentagon felt that "properly armed" military analysts could "push back" reporting
that contradicted the narrative it was promoting—for example, information that
troops were dying in Iraq because they did not have body armor. The Pentagon
closely monitored what these "properly armed" analysts said each time they
were on the air and responded swiftly if they deviated from the official narrative; several analysts told Barstow that they received an angry phone call from
Pentagon officials moments after they got off the air. One Fox News Channel
analyst felt that any criticism he might voice would lead to swift consequences

in the form of cancelled contracts. Another was actually dismissed for uttering a critical phrase on *The O'Reilly Factor.*

CNN pleaded ignorance regarding its analysts' business interests, and Fox News Channel did not comment on the story (Barstow 2008). On *The Bryant Park Project,* however, NPR confronted its own complicity in this charade (the network had drawn on Robert Scales, a Fox News Channel analyst, for analysis of the war). The hosts of the program also discussed the lack of coverage that the story had received on the news networks. Mike Pesca counted the number of times that the networks mentioned this latest government scandal. Only ABC, NBC, MSNBC, and CNN had dealt with the story, and CNN was the only network to mention it more than eight times. Scott Collins, a columnist for the *Los Angeles Times*, argued that the lack of coverage, as well as the lack of outcry, pointed to the public's acknowledgment that the "distinctions between the government and journalism" have been obliterated (National Public Radio 2008). The view is pessimistic, to be sure, but it is not hyperbolic. And it raises the question of where the public was at the height of the patriotic coverage.

## High-Concept News and the Audience

For high-concept texts to survive, there must be an audience willing to consume the primary text and the ancillary merchandise. Ratings and sales figures indicate that such an audience existed for the first year of the war coverage. Whether or not that audience bought into the ideological construction of the war coverage is unknown, but it is an important issue to consider. One study of the 2003 war coverage found that viewers' perceptions of the war varied according to their news sources. The study singled out Fox News Channel as an exceptionally uninformative source, although it acknowledged problems across most of the news networks.

"Misperceptions, the Media, and the Iraq War" (2003–2004) by Steven Kull, Clay Ramsay, and Evan Lewis, received a great deal of attention for its findings regarding Fox News viewers.[2] The authors asked why support for the war in

---

2. When the authors narrowed the field by networks (ABC, CBS, NBC, CNN, Fox, PBS, and NPR), they neglected to differentiate in their poll between news programs originating from the networks and news programs originating from local affiliates. For this reason, I will refer only to Fox and not specifically to Fox News Channel, as Fox affiliates are not the same entity as Fox News Channel. We also have no way of knowing

Iraq persisted even after no WMDs were found and no evidence was found to support a link between Iraq and Al-Qaeda. They polled viewers in order to understand how misperceptions about the war arose and who held those misperceptions. They identified three misperceptions: WMDs had been located, the United States had presented credible evidence to link Iraq to Al-Qaeda, and most of the world supported a U.S. invasion of Iraq. The findings, based on the polling of 3,334 respondents, provide startling information about what the U.S. public believes about the war. I will focus on one of their hypotheses regarding the influence of news sources on supporters of the war.

In the study, 520 respondents identified Fox as their main source of news, 466 identified CNN, and 91 identified NPR/PBS.[3] Of the 1,362 respondents questioned about all three misperceptions, 80 percent of Fox viewers had one or more misperception; the corresponding percentage for CNN viewers was 55 percent and for consumers of NPR/PBS only 23 percent. Sixty-seven percent of Fox viewers thought evidence of a clear link existed between Iraq and Al-Qaeda; the corresponding percent for CNN viewers was 48 and for NPR/PBS viewers and listeners only 16 percent. Thirty-three percent of Fox viewers believed that the United States had found WMDs, while 20 percent of CNN viewers and 11 percent of NPR/PBS viewers/listeners did. Thirty-five percent of Fox viewers felt that most of the world supported the war, while 24 percent of CNN viewers and only 5 percent of NPR/PBS viewers/listeners held that view (Kull, Ramsay, and Lewis 2003–2004, 582–584).

Kull, Ramsay, and Lewis wondered if perhaps the level of viewers' attention to the news had some connection to the polling results. Of the respondents who claimed they followed Fox "very closely," a staggering 80 percent believed that the United States had established a credible link between Iraq and Al-Qaeda (586). Among that same sample, 44 percent believed the United States had found WMDs, and 48 percent believed that most of the world supported the war effort. Considering that respondents who claimed not to follow the news at all had significantly lower percentages in these categories, the findings are telling. The authors provide no numbers for CNN viewers, but they found that among viewers who followed CNN very closely, there were "slightly, but significantly, lower

---

if or to what degree the networks pass their ideological or political identification on to the affiliates. The authors acknowledge this problem, but in spite of the complications, the findings indicate a dramatic problem within news networks.

3. I am including NPR/PBS for the sake of comparison.

levels of misperception" regarding WMDs and international support for the war but not for the link between Iraq and Al-Qaeda (ibid.). For the most part, the authors found no connection between how closely the respondents paid attention to news and the degree of misperception. Only in the case of Fox did a higher level of attention modestly increase the likelihood of misperceptions.

Kull, Ramsay, and Lewis offer several explanations for the high percentages of viewers who had misperceptions about the war. One explanation posits that anyone who listened to President Bush's statements from his October 28, 2003, press conference could have misinterpreted them at a number of points. A second explanation involves the "way that the media reported the news." The authors found evidence that reporting impacted misperceptions under certain circumstances and that different media outlets felt they should not "challenge" the White House. They also found that stories about possible or impending finds of WMDs were plentiful but that stories about the failure to find WMDs were rare. Kull, Ramsay, and Lewis argue that the focus on the constructed notion of "French obstructionism" possibly overshadowed global anti-war sentiment to create misperceptions about the level of international support for the war (592, 594).

The findings of the study indicate that Fox viewers were much more likely to have misperceptions about the war than CNN viewers. However, on some issues, the margins that separate CNN and Fox viewers are not that wide. If we look back at the percentages of the two networks' viewers who held misperceptions, we can see some variation in the gaps between them. On the high end, twenty-five percentage points separated Fox and CNN viewers, but on the low end, only thirteen percentage points separated Fox and CNN viewers. It seems the researchers were not preoccupied with the fact that both networks generated misperceptions. The study positions Fox as opinionated and slanted, while CNN maintains its brand of objectivity and neutrality simply by not being as opinionated or slanted as Fox. Nevertheless, the study is significant because it quantifies the degree to which viewers of the two networks were misinformed.

Many scholars are concerned that studies like this one provide evidence of the threat to democracy that conglomerate-controlled news poses. Kull, Ramsay, and Lewis' study is one step toward substantiating claims of outright disinformation. I have argued that the similarities between CNN and Fox News Channel tend to outnumber the stark differences, and the similarities substantially deflate Fox News Channel's argument that all other channels lean toward the left in their analysis of news. So, if Fox News Channel was created as an alternative to CNN, and both networks maintained the same narrative in the

*Conclusion*

first days of the 2003 invasion (and both generated high percentages of misinformed viewers), we can look to those elements specific to media—high concept and the genre of television news—to provide some insight into the similarities between these two commercial news outlets.

## High Concept and Media Studies

High concept began in television, migrated to film, and returned home again; the term is more than just a descriptor of post-classical Hollywood. It is a strategic model formed within the media industries and specific to them. Certainly, genres, stars, character types, and easily sold stories existed before high concept. However, when those aspects of U.S. filmmaking met the increasingly corporatized Hollywood of the 1980s, high concept as a concrete industrial formula was born. High concept became a way to contextualize a phase of Hollywood that continues to mature as conglomeration meets new media, new marketing channels, and new distribution windows.

One of the ways that high concept is invaluable to the study of commercial media is its specificity to commercial media. It is not an idea that gestated in literary studies and leapt to film studies. And it is not an idea that is particular to film and untranslatable to television. Cultivated and implemented by the minds that wanted to sell, sell easily, and sell big, high concept is a model that belongs to commercial media, even as they are in a constant state of textual, technological, and regulatory flux. As long as there are corporate media, high concept will exist in one form or another, as Wyatt implies (1994, 202). Although the continued overshadowing of art by commerce is certainly not a positive situation, our ability to trace this imbalance over time and across media historically and critically *is* positive. High concept offers one framework for doing so, and for better or for worse, it explains many developments in media since its inception.

In television, one of these developments is the large role that product placement and ancillary merchandise play in sustaining programs and, indeed, genres. One can point easily to DVD sales, episode downloads, soundtracks, and video games based on continuing character series. Aggressive marketing campaigns for specific programs on television, the Internet, in movie theaters, and in print are also commonplace. But just as some movies are high concept and some are not, some television programs lend themselves to the tenets of high concept, and some outwardly resist them. Children's programming formed a healthy relationship with merchandising in the days of radio. Now it has an unbreakable marriage with the logic of high concept. Another genre that adheres quite faithfully to high concept currently is "reality" programming. From talent programs like *American Idol* (Fox), *Top Chef* (Bravo), *Project Runway* (Bravo), *So You Think You Can Dance* (Fox), and *Dancing with the Stars* (ABC) to physical competition programs like *The Amazing Race* and *Survivor* (CBS) to faux-vérité celebrity programs like *The Osbournes* (MTV), *The Simple Life* (MTV), *Keeping up with the Kardashians* (E!), and *Denise Richards: It's Complicated* (E!), the replication of formulas seems to have outpaced scripted series.

Certainly, scripted series have not abandoned replication. One of the great examples at the moment is the forensic science series made enormously popular by Jerry Bruckheimer's *CSI* franchise—*CSI* (CBS), *CSI: Miami* (CBS), *CSI: New York* (CBS)—and continued on series like *Bones* (CBS) and *Dexter* (Showtime). Interestingly, these programs were preceded by and have existed alongside their reality-based brethren on basic cable—*The New Detectives* (TLC) and *Forensic Files* (truTV)—and pay cable—*Autopsy* (HBO).

Genres like cop, doc, and lawyer dramas have been and will continue to be omnipresent on television, but as programs like *The Wire* (HBO), *ER* (NBC), and *Damages* (FX) have updated and complicated narrative and generic conventions, we have also seen the family drama turned on its head by *The Sopranos* (HBO), *Dirty, Sexy, Money* (ABC), *Six Feet Under* (HBO), and *Big Love* (HBO). Additionally, programs like *Lost* (ABC), *Mad Men* (AMC), *Breaking Bad* (AMC), *Rome* (HBO), *Deadwood* (HBO), and *The Sopranos* (HBO) have tweaked a number of traditionally cinematic genres and dead television genres to push boundaries a bit further. Likewise, the turn to single-camera aesthetics and painfully awkward characters on sitcoms such as *Arrested Development* (FOX), *The Office* (NBC), *30 Rock* (NBC), *Extras* (HBO), *Curb Your Enthusiasm* (HBO), and *It's Always Sunny in Philadelphia* (FX) have moved the genre away from the static form of the multi-camera setup and toward a style heavily dependent on quick editing for comedic effect.

At the end of the programming day, many (but certainly not all) scripted series have rejected simplicity of character and narrative, while reality programs have embraced it. Reality TV's reliance on simple premises rivals their dependence on high-angle crane shots and excessively dramatic music and lighting. The celebrity-driven shows are simpler stylistically, but the pre-sold property eliminates the need for visual embellishment. In all cases, the programs are much cheaper to produce than scripted series, and some of the programs make efficient and blatant use of product placement. In one episode of *Project Runway*, we learn the benefits of L'Oréal makeup, TRESemmé hair products, Bluefly.com accessories, *Elle* magazine, and the American Express card. A 2008 article on the escalating use of product placement on television highlights the fruitful relationship between reality TV and marketing:

> Reality shows can—and do—get away with far more egregious pitching [than scripted series]: What is *Celebrity Apprentice* but QVC with occasional bickering (and an actual ad for QVC)? It's possible that a steady diet of plug-stuffed reality TV has numbed consumers to the trend's shift into scripted programming. (Armstrong 2008, 40)

If we consider the relatively low cost of production, the decent-to-high ratings, the seemingly endless marketing opportunities, and the insistence on style over substance, then we can point to reality programming as the most blatantly market-driven kind of television out there. These are just a few high-profile examples of how high concept has nestled into the TV landscape. The budgets may not be as inflated as Hollywood's cinematic offerings, but the adaptation and application of high concept to the specificities of television are apparent.

## Generic Coverage

Significantly, Armstrong's article concludes by citing the horror and inevitability of a news anchor "whet[ting] his whistle between stories with a can of Dr. Pepper," but news programs began with precisely this type of sponsorship (2008, 40). NBC's "experimental" news program in 1940, *The Esso Television Reporter,* was presented by NBC and Standard Oil, and its nightly news program in the early 1950s was *Camel News Caravan,* sponsored by Camel Cigarettes (Karnick 2003). The great fear now is that "a news sponsorship could be construed as an exertion of influence over the judgment of journalists" (Levingston 2005), but would we rather see the money exchange hands or naively presume that no entity exerts influence over news content at all?

Furthermore, if the genre of reality programming is conducive to the model of high concept, then how can we address the connection between news and high concept through a consideration of genre? Both Robin Andersen (2006) and Henry Giroux (2006b) write that coverage of the current war with Iraq exploited the production values of the reality program. Elayne Rapping (1987) and James Wittebols compare television news to soap opera. And I have already recounted scholars' and reporters' comparisons of war coverage with action films and video games. Why is TV news everything but TV news? Why do scholars resist acknowledging that TV news is its own genre, firmly planted in the television industry?

Genres shift over time in a cultural, political, and economic process of rein-vention. Jason Mittell asks us to reconcile these shifts with a degree of coherence that genres must maintain in order to have any meaning at all. "At any moment," Mittell writes, "a genre might appear quite stable, static, and bounded; however that same genre might operate differently in another historical or cultural con-text" (2004, 176). Those contexts transcend individual programs and include input from both the industry and the audience. As a genre operating within discursive practices and susceptible to economic pressures, television news should be no different from any other televisual genre.

For example, television news began as an offspring of the cinematic news-reel and radio news programming. The reliance on filmed footage initially meant waiting days for images. The adoption of satellite and video technology injected a renewed sense of "liveness" to the genre, which in turn transformed the role of television journalists. Similarly, the expansion of the news from fifteen minutes to thirty minutes in the 1960s and finally to twenty-four hours in the 1980s impacted more traditional notions of how the genre functioned. Cutbacks meant that network news could no longer survive on poor ratings, so changes in form and the type of news covered (the decline of international news in favor of domestic news) ensued. Television news has developed as all genres have developed. Furthermore, to argue that television news borrows its traits from reality TV is to discount the fact that news was the first form of "reality TV." It also dismisses the coherence of the genre of television news or sense-lessly attempts to confine news to one function of the genre: the dissemination of information.

The best signifier of a coherent genre is parody. *Saturday Night Live*'s "Weekend Update" began by spoofing the established style of network news in the 1970s and early 1980s. No program has parodied the generic traits of contemporary television news more effectively than *The Daily Show with Jon Stewart.* Billed as "fake news," but somehow more informative than CNN and Fox News Channel put together, *The Daily Show* nails the conventions of TV news with tremendous accuracy. From taped reports, man-on-the-street inter-views, and "remote" reports to graphics, music, correspondents, pundits, and guest experts, the late-night Comedy Central program solidifies what it means to be a news program right now. It also questions the authenticity ascribed to the genre by tampering with our assumptions about reality and photography. *The Daily Show* has achieved such a central position in our understanding of how news produces meaning that clips from the program make regular appear-ances on cable news. CNN International even carries *The Daily Show: Global Edition,* a weekly program that highlights the best segments from the previous week's episodes. In recent years, politicians and academics have outnumbered celebrity guest appearances on the program. And we should not overlook a 2004 Annenberg survey, which revealed that regular viewers of *The Daily Show* "were better informed" than viewers of network news were (Hadar 2007). The result is a constant and compelling exchange of ideas and barbs between the established news institutions and a satirical news program. This exchange underscores the stability and flexibility of any television genre—news included. More important, it calls attention to the way that elements of high concept have settled into the contemporary definition of television news.

The relocation of journalism onto television signaled an entirely new type of journalism as much as it signaled an entirely new television genre. Efforts to make television news more televisual take place in the context of the real world of network economics. But does television news have a commitment to remain sparse—to look and sound like CSPAN or PBS's *NewsHour with Jim Lehrer*? Or is it possible to maintain the style of cable news and the calm demeanor of *NewsHour*? Laurie Ouellette and Justin Lewis (2004) write that one of PBS's failings is its resistance to merging successful commercial genres with its non-commercial sensibility. Will a U.S. audience ever see the style of CNN or even Fox News Channel combined with the decidedly more thoughtful news delivery that *NewsHour* boasts?

In May 2005, CNN aired a simulcast of CNN International's news program *Your World Today*. In a network press release, CNN's president Jonathan Klein applauded his decision to air the program: "The simulcast demonstrates the depth and reach of the entire CNN News Group and not just one domestic network. . . . We cover international news better than anyone. Why not give our U.S. viewers a real opportunity to see the type of stories they cannot see anywhere else?" (TimeWarner.com 2005). Thomas Lang of *Columbia Journalism Review* watched the simulcast and was "stunned" by the experience. He writes that the program was "one solid hour of journalism with teeth [that] avoided the Talking Head Ping Pong matches that plague cable news." In his conclusion, he poses the following questions to Klein: "Why has it taken this long? Did CNN all along think that American viewers, besotted on tales of runaway brides and Court TV, couldn't handle informed discussion concerning matters in Sudan, Syria, and North Korea?" (Lang 2005).

While Lang focuses on the empty talk of cable news, we must consider how style fits into the conversation. Style does not automatically rob something of substance; it simply makes a lack of substance easier to swallow. Because Al Jazeera's English-language channel is notoriously difficult to access in the United States and CNN treats its international audience much differently than its domestic audience, U.S. viewers are isolated from news that is both staid and stylized.

## The Reach of High-Concept News

With a few exceptions, my analysis focuses solely on U.S. media—an approach that is both an opportunity and a limitation. So let us combine the two for a brief moment. Does high-concept news exist only in the United States? The answer depends on several factors particular to each country: the industry's ability (or desire) to develop a model similar to high concept, the successful exportation of high concept from Hollywood, and/or the specific traditions of television news. For example, while some local newscasts in Mexico rely on puppets or men dressed as clowns to provide political commentary, the national newscasts tend to avoid those methods. One now-defunct network based in Mexico City, CNI, launched a left-leaning news program, *CNI Noticias,* in the late 1990s to compete with TV Azteca's and Televisa's more conservative news organizations. *CNI Noticias* made no claims of objective reporting and used canted angles, zooms, and quick editing in its taped and edited stories. Music also played an important role, as it often commented on the stories ironically. Although the set design and news frame were simple and uncluttered, the program's sensationalistic use of visuals and sounds in the packaging of its news stories rivaled and even exceeded those practices on Fox News Channel. Its status as a commercial news enterprise was also very transparent. In 2005, certain segments on the program were sponsored by Tequila Centinela and Corona. Unlike CNN and Fox News Channel, *CNI Noticias* was intensely critical of Mexican president Vicente Fox's conservative administration. The style and marketing of television news in other countries may mimic the current style of high-concept news in the United States, but if the networks do not promote politically conservative values, can they truly be high concept? And if CNI had not been shut down by Vicente Fox's government and TV Azteca, how might its reporting have looked if the next administration had been leftist? Without an in-depth analysis, these are questions best left to a future investigation. As my work here focuses on U.S. television news, one question I can address at this time is whether high-concept news is widespread domestically.

One way of answering this question is to consider whether some news events are more or less conducive to high concept. Twenty-four-hour news *is* conducive to high concept, as is focused coverage on broadcast news, but not all news programming can be high concept. CNN and Fox News Channel are not necessarily the most high-concept–oriented networks in the cable or broadcast news arena, but their visibility and success make them obvious examples of television news trends. Similarly, the hyperbolic style and content of Fox News Channel make the network such a pop cultural icon that it is the subject of ridicule and parody on *The Daily Show* and *The Colbert Report*. In spite of their relatively small audiences, CNN and Fox News Channel are trendsetters in television news, and even the broadcast networks mimic their style and strategies. For instance, ABC News adhered to high concept's superficial celebration of high technology by employing the same type of weapons fetishization as CNN and Fox News Channel. But the broadcast network's time limitations do not allow for the degree of in-depth narrativization that the 24-hour format does. Significantly, such time constraints may have forced even more simplification than what CNN and Fox News Channel offered.

Nevertheless, the simplified war narrative on CNN and Fox News Channel endured. Faced with a narrative that threatened to buckle at many points, the persistence of CNN's and Fox News Channel's journalists managed to contain those threats at the beginning of the war. The vulnerability of the war narrative indicated that the constant flow of *new* information hampered its cohesion, so the appeal of breaking news is something of a disadvantage for high-concept news. That CNN and Fox News Channel were able to paper over cracks in their war narrative speaks to their commitment to high-concept news on more than just the visual, aural, or narrative levels. Not all news coverage exemplifies that level of dedication to the practice of high concept or to maintaining the dominant political discourse in general.

News coverage of the tragedy on September 11, 2001, exhibited some high-concept traits. The coverage was lengthy and plagued by an information vacuum, yet the Bush administration told a story that the networks disseminated. As a result, the coverage adhered to a brand of nationalism mirrored in the first days of the 2003 war coverage (recall Fox News Channel's pioneering use of the U.S. flag in its network logo). The September 11 coverage even shared some of the same narrative and marketing elements with the 2003 coverage. Clear heroes and villains emerged and persisted in the years following the attacks. Those same villains—adjusted to incorporate Saddam Hussein—and victims formed part of the official justification for the 2003 war. Merchandise even followed the attacks, as I discussed in chapter 6, and on April 28, 2006, Universal Pictures released *United 93*, a narrativized version of the events on United Airlines Flight 93. For better or worse, September 11 became toyetic, but this was not the case at the time of another disaster. News of Hurricane Katrina, the most recent large-scale news event before the 2008 election, was possibly less comfortable to produce than September 11 because there was no foreign "other" to demonize and thus no easy nationalistic narrative to attract the support of the country.

When Hurricane Katrina devastated Louisiana and Mississippi in 2005, the crisis could have been recouped within the framework of a high-concept film. The occasion cried out for the classic scene of the hero putting himself in peril to save anyone in danger. Indeed, the rescue of citizens in distress could have been as dramatic as the footage of the Jessica Lynch rescue. However, the aftermath of Hurricane Katrina pitted the U.S. government against the disenfranchised and mostly African-American citizens of Louisiana. Even though Hollywood actors and the Hollywood fund-raising machine intervened (as with September 11), the natural disaster and the manmade injustices it revealed were not toyetic. Hurricane Katrina was a low-concept news event.

At the outset, despair and incredulity united much of the U.S. public in the face of what seemed to be an easily explainable tragedy. However, in this instance, the "other"—African-American Louisianans—had a voice. Unlike the Iraqi civilians in 2003 and beyond, these victims could share their feelings to some degree, thereby complicating the simple narrative of a low-lying city hit by a Category 4 hurricane. The villains alternated between the Federal Emergency Management Agency, the governor of Louisiana, the mayor of New Orleans, President Bush, and the citizens themselves (those who did not have the means to leave the city were painted as irresponsible by bureaucrats). The root of the bureaucrats' villainy was, among other things, the woefully inadequate maintenance of flood safeguards around New Orleans. Such negligence and its very visible and audible consequences did not allow for narrative simplicity.

Because a simple narrative was not sustainable, Katrina the coverage and Katrina the event resisted commodification but did not stave it off entirely. The networks promoted their coverage, some familiar contractors surfaced during reconstruction, and some corporations used their involvement in the rescue and relief efforts in their advertisements. Katrina was what the war in Iraq became once the narrative dissolved: a floundering story with no established narrative trajectory, lackluster marketability, and an increasingly disenchanted consumer base.

U.S. news stories with an international component might be more conducive to the practice of high concept simply because television journalists can construct clear us-versus-them narratives. Purely domestic news stories like school shootings, hate crimes, and the health care crisis necessitate a degree of introspection and self-criticism that is not as marketable as a surprise attack against the United States on its own soil. An analysis of domestic news stories and those involving the United States and other countries might yield significant findings about the persistence of high-concept news. In the meantime, we can look at our television screens right now to study how war coverage transitioned from high concept to low concept.

### Where Did the War Go?

In 2003, Steve McKee wrote an article for *Adweek* that recalled how in World War II, Hollywood celebrities had contributed to the war effort by serving in uniform. Writing shortly after PFC Jessica Lynch was rescued by U.S. troops, McKee wrote, "Today, instead of Hollywood helping to supply the war effort, I'm afraid the war effort is going to supply Hollywood" (McKee 2003). Indeed, the invasion of Iraq has inspired Hollywood producers to mediate the war in their own ways. Promotional material for the film *The Core* (John Amiel) in 2003 emphasized the use of the USS *Constellation* in the filming (Holson 2003). Steven Bochco created a poorly received drama for FX in 2005 entitled *Over There* about soldiers fighting in Iraq, and Jerry Bruckheimer developed *Profiles from the Front Line,* a 2003 reality show on ABC about soldiers fighting in Afghanistan.

The Jessica Lynch situation was unique, however, because her amnesia about what happened for part of the time while she was captured made her story fodder for a publicity machine. The U.S. media sensationalized her story even before any facts emerged. Later Lynch revealed that her rifle had jammed repeatedly and she had not been able to fire a single shot to defend herself. She insists that she was not a hero. Despite Lynch's insistence that her story should be limited to the facts that she knows, the media has portrayed her as a "female Rambo" and as a rape victim who was brutally beaten on her hospital bed in Iraq. (Lynch says that she was not tortured and was well treated in the Iraqi hospital.) Her story included many inconvenient details: her equipment was defective, her convoy got lost and drove into an ambush even though it had GPS technology, and Lynch actively resisted being cast in the role of hero. But none of this stopped the media from constructing a narrative that upheld the marketable concept of superior U.S. technology, heroic U.S. soldiers, and a military force that would prevail. The Jessica Lynch story is an example of how the Iraq war "supplied Hollywood" with material for its marketing machines. Her rescue was fodder for a publicity machine that did not end with news reports. Indeed, as the news reports on the war have dwindled, other genres have filled the gaps (Mulrine 2008).

The Project for Excellence in Journalism's 2008 report on the news media reveals that in 2007, the Iraq war occupied only 10 percent of news stories on

Fox, 16 percent on CNN, and 18 percent on MSNBC (Project for Excellence in Journalism 2008, "Cable TV" section). In all of cable news during 2007, the top ten stories, beginning with the most covered, were the 2008 election campaign, Iraq policy, immigration, war events, domestic terrorism, the Virginia Tech shooting, the death of Anna Nicole Smith, the U.S. attorneys who were fired, and the Valerie Plame story. Tom Rosenstiel, the director of the Project for Excellence in Journalism, explains why the war was relegated to the number-four spot: "It's a lot easier to cover [the war] as a political debate in Washington than to cover it on the ground in Iraq" (Perez-Peña 2007). One might also argue that it's a lot less easy to cover a war that has no narrative, marketable concept, or nationalistic propaganda to accompany it.

As it has departed from the news, the war has made the rounds on television and in film with inconsistent results. *Saving Jessica Lynch,* the made-for-TV movie based on Lynch's ordeal, aired on NBC in November 2003, seven months after the rescue story broke. At this time, the country was not as disenchanted with the war or the administration as it is now. Because the original high-concept text was the war coverage, this made-for-TV movie—just like video games and songs—was a piece of merchandise that attempted to sustain interest in the original war narrative. The November 9, 2003, broadcast of *Saving Jessica Lynch* attracted 14.9 million viewers, ranking it sixteenth among the top twenty over-the-air television programs for the week of November 3 through November 9 (Entertainment Weekly 2003). Sixteenth place may seem meager by television's standards, but considering that the November 2003 viewership for CNN and Fox News Channel was 849,000 and 1.2 million, respectively, the viewership for *Saving Jessica Lynch* was impressive (Project for Excellence in Journalism 2005, "Prime Time Cable News Viewership"). The film aired in spite of the controversy surrounding its subject and made a decent showing, so as a tie-in to the war narrative its value exceeded just the ratings.

In many ways, *Saving Jessica Lynch* crystallizes the meaning and aim of high concept. Its embodiment in the made-for-TV movie genre recalls the origins of high concept. The title's reference to the 1998 Steven Spielberg film, *Saving Private Ryan,* recalls not only high concept's development in the film world but also the intertextual value of World War II. And its presence on General Electric's network, NBC, signals high concept's return to television as a purveyor of merchandise for a televised war. The layers of intertext and meaning, bound up in a television genre that was born of the 25-word pitch, speak to the efficient packaging of a high-concept package. What matters to high-concept news, though, is that the film as a tie-in was a highly visible marketing strategy for the war and for the coverage of the war.

CNN and Fox News Channel treated Lynch's rescue with reverence in the beginning, even though military analysts on Fox News Channel had assured journalists that Delta force would not launch a rescue mission (Fox News Channel March 23). Trace Gallagher, reporting from Fort Bliss, warned Fox News Channel not to "expect a commando raid to go in there and try and rescue these POWs" (Fox News Channel March 24). Despite his self-assurance, a commando raid did follow, and CNN and Fox News Channel journalists willingly immersed themselves in the emotion and drama of the rescue. Paula Zahn called it a "daring nighttime raid" and characterized the scenario as "Saving Private Lynch," one of the first references to Spielberg's film (CNN April 2). The brief video of U.S. soldiers carrying Lynch to safety on a stretcher aired continuously. Aaron Brown downplayed the poor visual quality of the video by assuring viewers that "the story and rescue of Private Jessica Lynch packs an emotional punch that comes through in living color" (CNN April 2). CNN even prepared computer animation to simulate how the rescue might have happened (ibid.). And when former MIA Scott O'Grady appeared on *Hannity & Colmes* to discuss the rescue of Lynch, Sean Hannity stated, "I want to make it clear to our enemies that we have the will to find [the perpetrators of the ambush] and hunt 'em down" (Fox News Channel April 1). Hannity's threat encapsulated the emotion of network journalists and commentators, using the fictive "we" to unite on-air journalists and guests, viewers, and soldiers in Iraq.

The Project for Excellence in Journalism studied the development of the Jessica Lynch story and found that despite evidence that refuted the Pentagon's version of events, the mainstream news media were "more likely [to] latch on to the more sensational version of events" (2003). The trajectory proceeded as follows: On April 1, 2003, CENTCOM informed the press that Lynch had been rescued; on April 2, her "wounds" became "gunshot wounds"; on April 3, Lynch was constructed as the "female Rambo"; on April 4, conflicting stories about her wounds circulated; then "the Iraqi lawyer who saved her life" entered the story. On April 7, the original story grew in an even more sensational manner with tales of her mistreatment as a prisoner; on April 15, the "questioning of what actually happened" began; and on May 15, the "reconsideration of the story" grew when a BBC report and subsequent documentary exposed the mishandling of the story (Project for Excellence in Journalism 2003). The *Chicago Tribune, The NewsHour with Jim Lehrer, The Washington Post,* and CNN all ran stories revising the original account of Lynch's treatment and rescue, but the acknowledgment of hyperbole and possible propaganda in the war coverage did not inhibit the creation of a product that exploited the original Lynch narrative. And with that, creative treatments of the war shifted dramatically.

Susan Carruthers concludes her book *The Media at War* by surveying the ways that visual media represented wars after they had concluded. It is worth considering how film and television treat an ongoing war, particularly since Carruthers asserts that media help "both to disinter and to bury past wars" (2003, 245). What follows is an admittedly brief look at the cinematic and tele-visual representations of the Iraq war.

Iraq has been incorporated into the storylines of too many made-for-TV movies to name here. As a central figure, though, the war has jumped from genre to genre and from medium to medium. Numerous documentary films have dissected the causes and consequences of the war in Iraq and the larger war on terror. One—Alex Gibney's *Taxi to the Dark Side*—earned an Academy Award in 2007. Michael Moore's 2004 anti-Bush, anti-war documentary *Fahren-heit 9/11* was the war's first big-screen outing, and it was controversial enough to earn over $119 million domestically. Another 2004 documentary of note that did not stir up the same amount of business as Moore's film was Michael Tucker and Petra Epperlein's *Gunner Palace*. Focused firmly on the soldiers in Baghdad, the film's tagline—"Some War Stories Will Never Make the Nightly News"—points to the disconnect between experience and representation.

In 2005, Steven Bochco created *Over There*, a series that attempted to pres-ent the ordeals of ground troops while eschewing a political bias. While Moore depended on politics to sell his film, Bochco tried to avoid taking any stance. The hyperviolent series failed to draw an adequate number of viewers, how-ever, and FX opted not to renew it for a second season. In 2006, *Home of the Brave*, Irwin Winkler's film about veterans returning from Iraq, earned just over $50,000 at the box office. On television, the war was domesticated in Lifetime's 2007 series *Army Wives*, a popular prime-time soap opera about the wives of deployed servicemen. Also in 2007, Brian DePalma's film *Redacted* pulled in only $65,000 in theaters. Robert Redford tackled the war on terror, as well, with his fiction film *Lions for Lambs*. It earned only $15 million in spite of an all-star cast (Strauss 2008). Also lightly attended was *Stop-Loss*, Kimberly Peirce's 2008 film about overworked soldiers. Tucker and Epperlein followed *Gunner Palace* with another documentary film in 2008, *Bulletproof Salesman*, a calculating look at a man who has amassed a small fortune selling armored vehicles in the war zones created by the United States. With the unpredictable performance of fiction films, documentaries, and television series, showrunners and filmmak-ers have learned that representing an ongoing war carries a degree of difficulty that has repelled cable news.

Despite the case against producing war-related fiction, HBO programmed David Simon and Ed Burns's *Generation Kill* in July 2008. A seven-part mini-series touting itself as "The New Face of American War," *Generation Kill* is "a reminder, if nothing else, this is still happening," according to co-producer Simon (Strauss 2008). If it seems as though HBO has stepped in when the news has stepped out, consider another of Simon's comments: "War is not pristine. It can't be defined by pure evil or good" (Strauss 2008). The producers' decision to go low concept, thereby retaining the reputation they earned from *The Wire*, contrasts sharply with the initial impulse of cable news networks to make the war high concept. *Generation Kill* does not have the burden of airing on an advertiser-supported channel, and because it has the clout of *The Wire* behind it, perhaps the prized HBO viewers will reopen their living rooms to the war.

One made-for-TV movie, one prime-time drama, one prime-time soap opera, several fiction films, countless compelling documentaries, and one mini-series later, the war continues. Yet the conflict can barely attract the attention of the network that appeared to bend over backward for it in 2003. Making war high concept is a process that requires a government public relations department that can plan a year ahead of time. It also requires nationalistic fervor and a compliant mouthpiece. Everything that springs from it—spectacle, disinformation, grand stories, provocative characters, displays of firepower, theme music, and toys—easily overshadows failed policies, crippled countries, and dead bodies. In 2003, we saw a concerted effort by many different parties to sell a concept backed up by passionate rhetoric. Years later, all media are struggling to give voice to complexity and outrage.

# Appendix A

*Graphic Frames at CNN and Fox News Channel,*
*March 19–24, 2003*

## A. CNN's Frames

All of CNN's frames retained the logo, text crawl, and banner at the bottom. One frame used three boxes, a large box on screen left that contained a night-scope image of Baghdad and two smaller boxes stacked on screen right. The box on top had a nighttime image of Baghdad, and the bottom box had a talking head. A variation on this design reversed the placement of the large and small boxes. A second frame had four boxes in total, the largest taking up three-quarters of screen right with a live shot of Baghdad at dawn. The remaining three boxes were stacked on the left quarter of the screen. The top box had a live shot of Kuwait City, the middle box had a live shot of Baghdad, and the bottom had a live videophone shot of a tank rolling through the desert. A variation on this frame included talking heads in the three smaller boxes commenting on the action in the larger box. In the third frame, four equally sized boxes took up the entire screen. This frame appeared most often during the bombings to display feeds from four different cameras in Baghdad.

## B. CNN's Thematic Frames

One thematic frame served as an update frame that contained assessments of different scenarios, like damage on the Iraqi side or the U.S. military divisions involved in the invasion. In this type of frame, screen left featured the *Strike on Iraq* logo superimposed over the right side of the golden Iraq graphic. The text underneath the logo communicated the title of the frame, and screen right contained descriptive text.

The "Latest Developments" frame consisted of a gold background that gradually became red at the center of the screen. Deep blue permeated the corners of the frame. The title appeared at the top, with the *Strike on Iraq* logo at screen right and the relevant text centered below the title.

The "Coming Up" frame was a split screen. The anchor was at screen right, and at screen left was a graphic of three file folders, each with upcoming topics in the coverage.

The "War Recap" frame had that title at screen left accompanied by an image box on screen right. The image changed with the information below it.

A frame dedicated to the embedded reporters and correspondents in the field included the *Strike on Iraq* logo at the top of the screen. At the center of the screen were three boxes with still photos of three reporters, and underneath the photos were their names and locations. The background was deep blue and red, with red occupying most of the space.

Finally, the "Battle Scenes" frame had a deep blue and deep red background. In the upper left-hand corner of the screen was a box with a glowing green animated tank. In golden letters was the title "Battle Scenes." On screen right was a box with footage. At one point, the footage portrayed a daytime shot of Baghdad with a plume of smoke rising.

### C. Fox News Channel's Frames

One of Fox News Channel's news frames consisted of two talking heads surrounded by boxes with folder-like tabs. Sharp diagonal beams splintered the blue-green background. Another similar frame contained two boxes with images inside them, but the boxes were shaped like narrow television screens. A third variation had two square boxes positioned diagonally from the top of screen left to the bottom of screen right, with the top left box overlapping the bottom right box. A three-box frame similar to CNN's had the large box on screen right and two smaller boxes with talking heads stacked on screen left. All boxes had the same television screen shape, and the background was red, blue, and green. Another three-box frame strayed from the visual evenness of the previous frames. Three boxes were arranged so that their formation resembled an upside-down V. Each box contained a talking head, and the text below read "Team Fox Coverage." A separate frame continued with this design, with a fourth box at the bottom of the screen so that the boxes formed a diamond.

### D. Fox News Channel's Thematic Frames

The network's first thematic frame during the first week of its wartime coverage was a general update frame. A map of the relevant area, in one instance a map of Iraq and Kuwait, appeared on screen left. Above the map was the word "Operation" with a colon after it. On screen right was a graphic of a blue globe against a black background. Placed over the globe was the text "Iraqi Freedom" (to follow the colon after "Operation") and the relevant topic, in this case "Helicopter Crashes."

The second major thematic frame, also geared toward updates, was entitled "War on Terror." In this frame, the outline resembled a file folder with a tab at the top but with another odd protrusion on screen left. The file folder appeared on screen by rotating into it. At the top was the title "War on Terror," and the outline of the file folder was filled with footage of recent events.

# Appendix B

*War-Related Animation Sequences
at CNN and Fox News Channel,
March 19–24, 2003*

### CNN's Cruise Missile Animation

The animated flight of a cruise missile aired early on March 20, 2003, after the first strikes intended to take out the Iraqi leadership and before the "shock and awe" bombing campaign commenced. At the beginning of the sequence, a graphic of the planet is visible from space, with an outline of the Middle East at the center of the frame. The text reads "Iraq" and "Baghdad"; Iraq is in orange, and Baghdad is under a red pushpin. In each successive frame, the map of Iraq appears larger. The text disappears as we zoom in to the region. Eventually we reach an immense warship in the Persian Gulf. We zoom in from a long shot of the ship, which has a cruise missile standing vertically atop it. We zoom to a closer shot of the missile launching and moving skyward. We follow the missile over the Gulf, past Kuwait, and into Iraq. The missile then flies into the word "Baghdad" and disappears into an inevitable explosion. Finally, we zoom back out to the initial establishing shot of planet earth.

### CNN's Bunker-Buster Animation

The sequence begins with a side view of a Stealth Fighter aircraft occupying the top half of the screen. The cargo doors open and the plane drops a bomb. After a cut, the bomb travels downward and we follow it as though we are attached to the tail end. As the bomb falls, the ground appears to get closer and the target finally comes into view. Another cut shifts the image to an overhead shot of a warehouse. The next shot is from the bomb's point of view as it rapidly approaches the warehouse. The next shot reveals a cross-section of the warehouse—it has four levels with assorted barrels and boxes scattered around each level. The missile enters the frame from the upper right-hand corner of the screen and penetrates each level at an angle. An explosion at the third level triggers three smaller explosions on levels one and two. One of the explosions produces black smoke, while several others produce green smoke, indicating the presence of chemicals. The resulting fire quickly disappears, leaving the charred remains of the building.

### Fox News Channel's Patriot-Scud Animation

This animated sequence begins with a shot of large sand dunes that have contours and texture. On the sand sit three Iraqi mobile Scud missile launchers aligned diagonally from the upper left to the lower right of the screen. The start of the scene is swift; we immediately begin to zoom out from the launchers as soon as they appear. The first launcher is hit by a Patriot and explodes as it leans back to launch its missile. The other two launch their Scuds sequentially and successfully. We continue to zoom out when the Scuds launch, following them as they fly upward in the direction of screen left. A Patriot intercepts the second missile in mid-air. The third, after reaching a certain height, turns downward. At the last second, a Patriot emerges from the bottom of screen right and intercepts the Scud before it hits its target.

# Appendix C

*Promotional Spots at CNN and Fox News Channel,*
*March 19–24, 2003*

### Promotional Spot at CNN

This promotional spot touted CNN's embedded reporters and aired during prime time on March 22, 2003. The voice-over, detailed in Chapter 6, matched the following visuals: A CNN logo appears at the center of the frame, then the promo cuts to a box outlined in white taking up five-sixths of the right side of the frame. Inside it is Martin Savidge speaking in front of a camera next to a soldier sitting by a military vehicle. To the left of the box is a deep red hue mixed with black and some unidentifiable images outlined in black.

Text intermingles with the visuals to reinforce the spoken words. The text reads "ALONGSIDE." This text disappears and two other words—"THE TROOPS"—shoot into the box. As this image transitions to the next, the background reveals an outline of the Middle East and specifically Iraq. The spot cuts to a box that is more centered than the previous so that the red background is more apparent on either side of it. Inside it is footage of a soldier loading a bomb. The text reads "WITH SOLDIERS." Another cut takes us to a box with footage of Walter Rodgers reporting in front of military vehicles. The text reads "WALTER RODGERS." Another cut takes us to a box with footage of Dr. Sanjay Gupta in a flak jacket standing in front of stacks of sandbags. The text reads "DR. SANJAY GUPTA." A fourth cut takes us to a box with footage of Ryan Chilcote standing in a blue polo shirt in front of barbed wire. The text reads "RYAN CHILCOTE."

The promo then makes an elaborate transition in which the white outlines of the box collapse vertically and re-expand to reveal another box positioned slightly to the left of the previous one. The red background is still visible on either side of the box. For the segments showing sailors, airmen, and marines, the same visual elements apply.

Footage unrelated to embedded reporters consists of a shot of two soldiers pushing an inflatable raft onto a ramp (for the "With sailors" voice-over), a shot from the point of view of a pilot in a cockpit (for the "With airmen" voice-over), and a shot of soldiers walking into a room (for the "With marines" voice-over).

When the voice-over is finished recounting the names of the embedded journalists, there is a shot of a reporter in front of a helicopter with the U.S. flag waving in the background. The on-screen text is the *TIME Magazine* logo. The promo cuts to a box with soldiers peeking out of the top of a tank, and the accompanying text is the logo for *The New York Times*. It then cuts to a box with a soldier opening the cockpit door of a plane, and *The New York Times* logo is replaced by the logo for the *Boston Globe*. It cuts again to a box with a soft image of a soldier in close-up. The *Atlanta Journal-Constitution* logo replaces the *Boston Globe* logo.

More shots of the embedded journalists follow, and the action picks up at the conclusion of the promo when the voice-over announces "No one gets you closer than CNN." Cut to a box with a soldier running, his back to the camera. A cameraman (with another camera trained on him) follows the soldier.

### Promotional Spot at Fox News Channel

Fox News Channel aired a 30-second promotional spot comparable to CNN's but with considerably more cuts and a faster pace. The voice-over is detailed in chapter 6. The visuals in the promo are an extension of the overall look of Fox News Channel. The primary colors are silver, blue, and white. The first image is one that reappears in the promo whenever "Fox News Channel" is spoken; it is the official image of Fox News Channel. A box outlined in silver and blue opens up at the center of the frame with the Fox News Channel logo located in the middle and "Fox News Channel" written in smaller typeface at the lower left corner of the box. Below the box, the Fox News Channel logo appears upside down and backward. The box disappears and is replaced with a barely visible grid pattern that occupies the entire screen. Three boxes of footage appear at three corners of the frame, with the Fox News Channel logo still at the center of the screen. The frame elements are in constant motion, and this is the case with every subsequent frame.

All the elements disappear in a flash of white light, and there is a cut to a new image with footage of soldiers marching in the background. A large circle with crosshairs materializes, and another flash of white light reveals the *Operation Iraqi Freedom* logo. The entire frame becomes enveloped in the crosshairs circle, and another cut transitions the sequence to a new series of shots of embedded reporters and reporters in the field.

Boxes with the images of the embedded journalists, which appear sequentially as in the CNN promo, shift around the screen. Accompanying their images are thick vertical fluorescent green and blue bars. Text that is initially composed of random letters unscrambles to reveal the name of each reporter. Smaller, unreadable text resembling streams of data on a computer screen appears in random areas. Boxes with footage of planes and other military-related actions accompany the images of the moving boxes with journalists' images.

When the parade of reporters stops, a cut returns the sequence to the initial Fox News Channel logo frames. What follows is a series of images devoted solely to the military. The first image is a full-frame shot from a tank's point of view. The camera is positioned behind the tank's gun. The text "On the Ground" appears in a black box with a horizontal golden beam. The next image is the reverse of the previous point-of-view shot. The camera is now facing a soldier behind a tank's gun head-on. The promo cuts to the tank's point of view once more. The tank fires, and the next shot reveals a golden explosion. Following that is a canted shot of a tank rolling from the bottom left of the frame to the top right. Then it cuts to an airborne pilot's point of view. As the camera shows a view from the cockpit, a stream of illegible data appears in a column on the right side of the screen. The text "In the Air" appears and the screen cuts to a long shot of a jet flying overhead. Another cut shows a shot of soldiers' legs running in the desert, followed by a wider shot of these soldiers in action. The text "From the Front" appears. The shots that follow include a soldier loading what looks to be ammunition, another soldier hunched over and walking away from a jet taking off from a carrier, the jet taking off, and a group of soldiers pointing guns in the desert. Next is an ultra-fast series of shots featuring a close-up of a soldier with a helmet on, a soldier looking intently into a viewer, and an exterior shot of a tank firing. The progression of the cuts implies that the soldier was loading ammunition, aiming, and firing the tank's gun. The military images conclude with a shot of a jet flying farther away from the aircraft carrier. The initial Fox News Channel logo frame appears and wraps up that sequence.

The next sequence highlights the political aspects of the war, with shots of President Bush at a podium, Ari Fleischer at a podium, and a very low-angle shot of General Myers at a podium.

The promo concludes by focusing on the embedded journalists. A camera in the desert is trained on a Hummer driving toward it. The text "The First Place to Turn" materializes letter by letter. The Hummer passes by the camera, and the video slows so that the viewer can see the civilian driver. The text "The Latest News" appears as the Hummer passes by the camera. The initial Fox News Channel logo frames appear again, this time with yet another logo frame bearing the tagline: "Real Journalism. Fair and Balanced."

# Works Cited

Agence France-Presse. 2003. "World Condemns Iraq War amid Fears of Humanitarian Disaster." March 20. Retrieved from LexisNexis Database.

Al Jazeera. 2008. "About Us: Corporate Profile." Retrieved from http://english.aljazeera.net/aboutus.

Altman, Rick. 1986. "Television/Sound." In *Studies in Entertainment: Critical Approaches to Mass Culture,* ed. Tania Modleski 39–54. Bloomington: Indiana University Press.

———. 1992. "Introduction: Four and a Half Film Fallacies." In *Sound Theory, Sound Practice,* ed. Rick Altman, 35–45. New York: Routledge.

Andersen, Robin. 1995. *Consumer Culture and TV Programming.* Boulder, Colo.: Westview Press.

———. 2006. *A Century of Media, a Century of War.* New York: Peter Lang Publishing.

Anderson, Christopher. 1994. *Hollywood TV: The Studio System in the Fifties.* Austin: University of Texas Press.

Anderson, Dante. 2003. "Kuma: War Developer Diary #2." *Action Vault,* November 11. Retrieved from http://www.leetforum.com/forumz/showthread.php?t=68101.

Anderson, Robert. 1985. "The Motion Picture Patents Company: A Reevaluation." In *The American Film Industry,* ed. Tino Balio, 133–152. Rev. ed. Madison: University of Wisconsin Press.

Armstrong, Jennifer. 2008. "Ad Nauseum." *Entertainment Weekly,* July 25.

Associated Press. 2003. "Winning This Fight: Some See Link between GI Joe Sales, War." *JSOnline: Milwaukee Journal Sentinel,* April 5. Retrieved from http://www.jsonline.com/news/gen/apr03/131441.asp.

Bagdikian, Ben H. 2000. *The Media Monopoly.* 6th ed. Boston: Beacon Press.

Balio, Tino. 1998. "'A Major Presence in all of the World's Important Markets': The Globalization of Hollywood in the 1990s." In *Contemporary Hollywood Cinema,* ed. Steve Neale and Murray Smith, 48–73. London and New York: Routledge.

Barstow, David. 2008. "Behind TV Analysts, Pentagon's Hidden Hand." *New York Times,* April 20.

Baudrillard, Jean. 1995. *The Gulf War Did Not Take Place.* Trans. Paul Patton. Bloomington: Indiana University Press.

BBC News UK Edition. 2003. "Sheen Leads 'Virtual' Iraq Protest." February 21. Retrieved from http://news.bbc.co.uk/1/hi/entertainment/showbiz/2787137.stm .

BBC News World Edition. 2003. "Actor Willis' $1M Saddam Bounty." September 26. Retrieved from http://news.bbc.co.uk/go/pro/fr/-/2/hi/entertainment/3141942.stm.

Belton, John. 1999. "Technology and the Aesthetics of Film Sound." In *Film Theory and Criticism: Introductory Readings,* ed. Leo Braudy and Marshall Cohen, 376–384. 5th ed. New York and Oxford: Oxford University Press.

Berkowitz, Bill. 2002. "War Toys for Tots." April 10. Retrieved from http://www.workingforchange.com/article.cfm?ItemID=13118 (accessed February 2004).

Berrigan, Frida, William D. Hartung, and Leslie Heffel. 2005. *U.S. Weapons at War 2005: Promoting Freedom of Fueling Conflict? US Military Aid and Arms Transfers since September 11.* A World Policy Institute Special Report. Retrieved from http://www .worldpolicy.org/projects/arms/reports/wawjune2005.html.

Bhatnagar, Parija. 2003. "Al-Jazeera Ousted from NYSE: Exchange Spurns Arab Satellite Channel, Network Regrets NYSE's Decision; Web Site Attacked." *CNNMoney.com,* March 25. Retrieved from http://money.CNN.com/2003/03/25/news/nyse_aljazeera.

Blumenstein, Rebecca, and Matthew Rose. 2003. "The Thesaurus Plays Its Part in the War Efforts." *San Diego Union-Tribune,* April 3. Retrieved from LexisNexis Database.

Bordwell, David. 1985. *Narration in the Fiction Film.* Madison: University of Wisconsin Press.

Bourdieu, Pierre. 1998. *On Television and Journalism.* Trans. Priscilla Parkhurst Ferguson. London: Pluto Press.

Browne, Nick. 1994. "The Political Economy of the Television (Super) Text." In *American Television: New Directions in History and Theory,* ed. Nick Browne, 69–80. Chur, Switzerland: Harwood Academic Publishers.

Burnett, John. 2003. "War Stories." *The Texas Observer,* June 6. Retrieved from http:// www.texasobserver.org/article.php?aid=1366.

Caldwell, John Thornton. 1995. *Televisuality: Style, Crisis, and Authority in American Television.* New Brunswick, N.J.: Rutgers University Press.

———. 2000. "Excessive Style: The Crisis of Network Television." In *Television: The Critical View,* ed. Horace Newcomb, 649–686. 6th ed. New York: Oxford University Press.

Carroll, Noël. 1995. "Towards an Ontology of the Moving Image." In *Philosophy of Film,* ed. Cynthia A. Freeland and Thomas E. Wartenberg. New York: Routledge.

Carruthers, Susan L. 2000. *The Media at War.* New York: Palgrave Macmillan.

Carter, Bill, with Jane Perlez. 2003. "A Nation at War: The Networks: Channels Struggle on Images of Captured and Slain Soldiers." *New York Times,* March 24.

Cavell, Stanley. 1979. *The World Viewed: Reflections on the Ontology of Film.* Cambridge and London: Harvard University Press.

CNN. 1994. "General Ralston Says Expanded Air Patrols Can Proceed." September 4. Transcript Retrieved from LexisNexis Database.

CNN.com. 2003a. "War in Iraq." Retrieved from http://edition.cnn.com/SPECIALS/2003/ iraq.

CNN.com. 2003b. "'Decapitation Strike' Was Aimed at Saddam." March 20. Retrieved from http://www.cnn.com/2003/WORLD/meast/03/20/sprj.irq.target.saddam.

CNN.com. 2005. "So-Called US Hostage Appears to be Toy." February 1. Retrieved from http://edition.cnn.com/2005/WORLD/meast/02/01/iraq.hostage.

CNN.com International Edition. 2003. "Commander-in-Chief Lands on USS Lincoln." May 2. Retrieved from http://edition.cnn.com/2003/ALLPOLITICS/05/01/ bush.carrier.landing.

Cohen, Richard M. 1997. "The Corporate Takeover of the News: Blunting the Sword." In Eric Barnouw et al., *Conglomerates and the Media,* 31–60. New York: New Press.

Collins, Jim. 1992. "Postmodernism and Television." In *Channels of Discourse, Reassembled,* ed. Robert C. Allen, 327–353. 2nd ed. Chapel Hill: University of North Carolina Press.

Collins, Scott. 2004. *Crazy Like a Fox: The Inside Story of How Fox News Beat CNN*. New York: Portfolio.

Cook, David A. 1996. *A History of Narrative Film*. 3rd ed. New York and London: W.W. Norton & Co.

Corner, John. 2002. "Sounds Real: Music and Documentary." *Popular Music* 21 (3): 357–366.

Debord, Guy. 1967. *Society of the Spectacle*. Repr., Detroit, Mich.: Black & Red, 1983.

Dellavigna, Stefano, and Ethan Kaplan. 2007. "The Fox News Effect: Media Bias and Voting." *Quarterly Journal of Economics* 122 (3): 1187–1234.

DemocracyNow.org. 2005. "Top Al-Jazeera Reporter Yousri Fouda on the Media and His Interviews with Al Qaeda leaders." October 13. Retrieved from http://www .democracynow.org/article.pl?sid=05/10/13/1359246&tid=25.

Denton, Robert E., Jr., ed. 1993. *The Media and the Persian Gulf War*. Westport, Conn.: Praeger.

Der Derian, James. 2001. *Virtuous War: Mapping the Military-Industrial-Media-Entertainment Network*. Boulder, Colo.: Westview Press.

Dobrin, Peter. 2003. "Media's War Music Carries a Message." *Philadelphia Inquirer*, March 23, A16.

Donovan, Robert J., and Ray Scherer. 1992. *Unsilent Revolution: Television News and American Public Life, 1948–1991*. Cambridge: Cambridge University Press.

Dorman, William. 2006. "A Debate Delayed Is a Debate Denied: U.S. News Media Before the 2003 War With Iraq." In *Leading To the 2003 Iraq War: The Global Media Debate*, ed., Alexander G. Nikolaev and Ernest A. Hakanen, 11–22. New York: Palgrave MacMillan.

DragonModelsLtd.com. 2003. "Dragon Extra 4: Operation Iraqi Freedom." Retrieved from http://www.dragonmodelsltd.com/catalog/extra_more.htm.

Dunn, Anne. 2003. "Telling the Story: Narrative and Radio News." *The Radio Journal: International Studies in Audio and Broadcast Media* 1 (2): 113–127.

Dyer, Richard. 1998. *Stars*. New ed. London: British Film Institute.

Eagleburger, Lawrence S., Claus Kleber, Steven Livingston, and Judy Woodruff. 2003. "The CNN Effect." In *The Media and the War on Terrorism*, ed. Stephen Hess and Marvin Kalb, 63–82. Washington, D.C.: The Brookings Institution.

El-Nawawy, Mohammed, and Adel Iskandar. 2002. *Al-Jazeera: How the Free Arab News Network Scooped the World and Changed the Middle East*. Cambridge, Mass.: Westview Press.

Ellin, Abby. 2003. "Two New Campaigns Reflect War News." *New York Times*, March 19.

Ellis, John. 1999. "Broadcast TV as Sound and Image." In *Film Theory and Criticism: Introductory Readings*, ed. Leo Braudy and Marshall Cohen, 385–394. 5th ed. New York and Oxford: Oxford University Press.

Engelhardt, Tom. 1994. "The Gulf War as Total Television." In *Seeing through the Media: The Persian Gulf War*, ed. Susan Jeffords and Lauren Rabinovitz, 81–96. Rutgers, N.J.: Rutgers University Press.

Engstrom, Nicholas. 2003. "The Soundtrack for War." *Columbia Journalism Review* 42 (1): 45–47.

Entertainment Weekly. 2003. "Get Smart." November 21. Retrieved from LexisNexis Database.

Fahmy, Shahira, and Thomas J. Johnson. 2007. "Show the Truth and Let the Audience Decide: A Web-based Survey Showing Support among Viewers of Al-Jazeera for Use of Graphic Imagery." *Journal of Broadcasting and Electronic Media* 51 (2): 245–264.

Fiske, John. 1987. *Television Culture.* London: Methuen & Co.

Fournier, Ron. 2003. "Bush Accuses Russian Firms of Aiding Iraq in War Effort." Associated Press, March 24. Retrieved from LexisNexis Database.

Friedman, Stefan C. 2003. "'Dirty Dozen' Bounty: U.S. Puts $1M Price on Saddam's Thug-ocracy." *New York Post,* December 28. Retrieved from LexisNexis Database.

Frontline. 1996. "The Gulf War: Weapons: MIM-104 Patriot." Retrieved from http://www.pbs.org/wgbh/pages/frontline/gulf/weapons/patriot.html.

Gans, Herbert J. 2003. *Democracy and the News.* Oxford: Oxford University Press.

Garamone, Jim. 1998. "U.S. Strikes Aimed at Iraqi Weapons of Mass Destruction." Armed Forces Press Service, December 17. Retrieved from http://www.defenselink.mil/news/newsarticle.aspx?id=41730.

Gaudiosi, John. 2003. "Kuma Plays War Games." *Hollywood Reporter.com,* August 14. Retrieved from http://www.hollywoodreporter.com/thr/new_media/article_display.jsp?vnu_content_id=1956066.

Gibson, Owen. 2007. "Murdoch Wants Sky News to be More Like Rightwing Fox." *The Guardian,* November 24. Retrieved from http://www.guardian.co.uk/media/2007/nov/24/bskyb.television.

Giroux, Henry A. 2006a. *Beyond the Spectacle of Terrorism: Global Uncertainty and the Challenge of the New Media.* Boulder, Colo.: Paradigm Publishers.

———. 2006b. *Stormy Weather: Katrina and the Politics of Disposability.* Boulder, Colo.: Paradigm Publishers.

GlobalSecurity.org. 2005. "Operation Northern Watch." Retrieved from www.globalsecurity.org/military/ops/northern_watch.htm (accessed January 2, 2008).

Glynn, Kevin. 2000. *Tabloid Culture: Trash Taste, Popular Power, and the Transformation of American Television.* Durham, N.C.: Duke University Press.

Gramsci, Antonio. 1997. *Selections from the Prison Notebooks.* Ed. and trans. Quentin Hoare and Geoffrey Nowell Smith. New York: International Publishers.

Greppi, Michele. 2003. "Adjusting to Costs of War." *Television Week,* April 7.

Hadar, Leon. 2007. "When Fake News and Comedy Trump Reality." *The Business Times* (Singapore), November 9. Retrieved from LexisNexis Database.

Hallin, Daniel C. 1986a. *The "Uncensored War": The Media and Vietnam.* Berkeley: University of California Press.

———. 1986b. "We Keep America on Top of the World." In *Watching Television,* ed. Todd Gitlin, 9–41. New York: Pantheon Books.

———. 1994. "Images of the Vietnam and the Persian Gulf Wars in U.S. Television." In *Seeing through the Media: The Persian Gulf War,* ed. Susan Jeffords and Lauren Rabinovitz, 45–58. Rutgers, N.J.: Rutgers University Press.

Hamilton, William L. 2003. "Toymakers Study Troops, and Vice Versa." *New York Times,* March 30.

Harper, Jennifer. 2003. "News Anchors Glum amid Iraqi Jubilation; Skeptical of U.S. Flags in Baghdad." *The Washington Times,* April 10. Retrieved from LexisNexis Database.

Herman, Edward, and Noam Chomsky. 2002. *Manufacturing Consent: The Political Economy of the Mass Media.* 2nd ed. New York: Pantheon Books.

Holson, Laura M. 2003. "The Media Business: Hollywood Toning Down Ads and Froth During War." *New York Times,* April 1.

Horwitz, Robert B. 1989. *The Irony of Regulatory Reform.* New York: Oxford Press.

Iraq Body Count. 2009. "Documented Civilian Deaths from Violence." Retrieved from http://www.iraqbodycount.org/database (accessed February 9, 2009).

Jacobs, Jill Rachel. 2003. "Coming to a Store Near You: GI George." *The Toronto Star,* August 25. Retrieved from LexisNexis Database.

Johnson, Chalmers. 2003. "Iraq Wars." *ZNet,* January 4. Retrieved from http://www.zmag.org/content/showarticle.cfm?ItemID=2869.

Johnson, Peter. 2007. "Media, Government Duel in 'Perfect Storm'; New Tension Brews in this Time of War." *USA Today,* February 13. Retrieved from LexisNexis Database.

Johnson-Cartee, Karen S. 2005. *News Narratives and News Framing: Constructing Political Reality.* Lanham, Md.: Rowman & Littlefield Publishers.

Karim, Karim H. 2003. *Islamic Peril: Media and Global Violence.* Montréal: Black Rose Books.

———. 2006. "American Media's Coverage of Muslims: The Historical Roots of Contemporary Portrayals." In *Muslims and the News Media,* ed., Elizabeth Poole and John E. Richardson, 116–127. London and New York: I. B. Tauris.

Karnick, Kristine Brunovska. 2003. "NBC and the Innovation of Television News, 1945–1953." In *Connections: A Broadcast History Reader,* ed. Michelle Hilmes, 85–99. Belmont, Calif.: Wadsworth Publishing.

Katzman, Kenneth. 2003. "Iraq: Weapons Programs, U.N. Requirements, and U.S. Policy." Issue brief for Congress. Retrieved from http://www.iwar.org.uk/news-archive/crs/19850.pdf.

Kellner, Douglas. 1992. *The Persian Gulf TV War.* Boulder, Colo.: Westview Press.

———. 2005. *Media Spectacle and the Crisis of Democracy: Terrorism, War, and Election Battles.* Boulder, Colo.: Paradigm Publishers.

Kirby, David. 1999. "When the News Drives the Music." *New York Times,* April 25.

Klaprat, Cathy. 1985. "The Star as Market Strategy: Bette Davis in Another Light." In *The American Film Industry,* ed. Tino Balio, 351–376. Rev. ed. Madison: University of Wisconsin Press.

Kolodzy, Janet, Wade S. Ricks, and Jill Rosen. 2003. "Stocks Are Down and Al Jazeera's Out." *American Journalism Review* 25, no. 4: 16.

Kosterman, Andrew. 2003. "Media Becomes Part of the Team in Deployments." *U.S. CENTCOM News,* March 8. Retrieved from http://www.au.af.mil/au/awc/awcgate/centcom/media_part_of_team.htm.

Kozloff, Sarah. 1992. "Narrative Theory and Television." In *Channels of Discourse, Reassembled,* ed. Robert C. Allen, 67–100. 2nd ed. Chapel Hill: University of North Carolina Press.

Kracauer, Siegfried. 1960. *Theory of Film: The Redemption of Physical Reality.* New York: Oxford University Press.

Krugman, Paul. 2007. "The Murdoch Factor." *New York Times,* June 29.

Kull, Steven, Clay Ramsay, and Evan Lewis. 2003–2004. "Misperceptions, the Media, and the Iraq War." *Political Science Quarterly* 118 (4): 569–598.

Kuma\War. 2008a. "Mission List." Retrieved from http://www.kumawar.com/Mission .php.

———. 2008b. "Mission List." Retrieved from http://www.kumawar.com/Mission .php.

Laing, Allan. 2003. "Operation Right Name Must Be Sound Decision." *Glasgow Herald,* March 12. Retrieved from LexisNexis Database.

Lang, Thomas. 2005. "CNN Stuns U.S. with Actual News."*Columbia Journalism Review,* June 6. Retrieved from http://www.cjr.org/behind_the_news/cnn_stuns_us_with_ actual_news.php.

Levingston, Steven. 2005. "Putting Their Names All Over the News: Banks' Sponsorship of Radio Newsrooms Raises Questions about Journalism Ethics." *Washingtonpost .com,* December 15. Retrieved from http://www.washingtonpost.com/wp-dyn/content /article/2005/12/14/AR2005121402225.html.

Londoner, David J. 1985. "The Changing Economics of Entertainment." In *The American Film Industry,* ed. Tino Balio, 603–630. Rev. ed. Madison: University of Wisconsin Press.

McAlister, Melani. 2005. *Epic Encounters: Culture, Media, and U.S. Interests in the Middle East since 1945.* Updated ed. Berkeley: University of California Press.

McChesney, Robert. 1999. *Rich Media, Poor Democracy: Communication Politics in Dubious Times.* New York: New Press.

McKee, Steve. 2003. "Heroes and Villains: Why Jessica Lynch Doesn't Belong on a Wheaties Box." *Adweek,* April 21. Retrieved from LexisNexis Database.

McLane, Brendan. 2004. "Reporting from the Sandstorm: An Appraisal of Embedding." *Parameters* (Spring): 77–88. Retrieved from http://carlisle-www.army.mil/usawc/ Parameters/04spring/mclane.htm (accessed August 2005).

McNair, Brian. 1998. *The Sociology of Journalism.* London: Arnold Publishers.

Merle, Renae. 2003. "Battlefield Is a Showcase for Defense Firms; Army Exporters Could Thrive on Televised Success in Iraq." *The Washington Post,* April 1.

Metz, Christian. 1999. "Aural Objects." In *Film Theory and Criticism: Introductory Readings,* ed. Leo Braudy and Marshall Cohen, 356–359. 5th ed. New York and Oxford: Oxford University Press.

Miracle, Tammy L. 2003. "The Army and Embedded Media." *Military Review* (September–October): 41–45. Retrieved from http://findarticles.com/p/articles/mi_m0PBZ/ is_5_83/ai_111573648.

Mittell, Jason. 2004. "A Cultural Approach to Television Genre Theory." In *The Television Studies Reader,* ed. Robert C. Allen and Annette Hill, 171–181. London: Routledge.

Moeller, Susan D. 1999. *Compassion Fatigue: How the Media Sell Disease, Famine, War, and Death.* New York: Routledge.

Morse, Margaret. 1986. "The Television News Personality and Credibility: Reflections on the News in Transition." In *Studies in Entertainment: Critical Approaches to Mass Culture,* ed. Tania Modleski. Bloomington: Indiana University Press.

Mullen, Megan. 2003. *The Rise of Cable Programming in the United States.* Austin: University of Texas Press.

Mulrine, Anna. 2008. "Former POW Jessica Lynch Recalls Her Captivity in Iraq." *U.S. News and World Report,* March 18. Retrieved from usnews.com.

National Cable & Telecommunications Association. 2008. "Top 25 MSOs—As of March 2008." http://www.ncta.com/Statistic/Statistic/Top25MSOs.aspx (accessed July 2008).

National Public Radio. 2008. "Networks Let Pentagon-Press Scoop Wither." April 24. Retrieved from http://www.npr.org/templates/story/story.php?storyId=89904775.

Newport, Frank. 2003. "Seventy-Two Percent of Americans Support War against Iraq." *Gallup Poll News Service,* March 24. Retrieved from http://www.gallup.com/poll/content/?ci=8038 (March 2005).

NOW on PBS. 2005. "Al-Jazeera and Arab Press Online." January 14. http://www.pbs.org/now/politics/aljazeera2.html.

NPD Group. 2003a. "Industry Trends: Top 10 Video Game Titles 1st Quarter 2003." Retrieved from http://www.npdfunworld.com/funServlet?nextpage=trend_body.html&content_id=395 (accessed February 2005).

———. 2003b. "Industry Trends: November 2003 Video Game Best-Selling Titles and Console Accessories." Retrieved from http://www.npdfunworld.com/funServlet?nextpage=trend_body.html&content_id=620 (accessed February 2005).

———. 2003c. "Industry Trends: December 2003 Video Game Best-Selling Titles and Console Accessories." Retrieved from http://www.npdfunworld.com/funServlet?nextpage=trend_body.html&content_id=781 (accessed February 2005).

———. 2004. "Video Game Sales Data for 2003." *About.com.* http://retailindustry.about.com/od/seg_toys/a/bl_npd012703_2.htm. In author's possession.

Ouellette, Laurie, and Justin Lewis. 2004. "Moving Beyond the 'Vast Wasteland': Cultural Policy and Television in the United States." In *The Television Studies Reader,* ed. Robert C. Allen and Annette Hill, 52–65. London: Routledge.

Parsons, Patrick R., and Robert M. Frieden. 1998. *The Cable and Satellite Television Industries.* Boston: Allyn & Bacon.

Pasquarett, Michael. 2003. "Reporters on the Ground: The Military and Media's Joint Experience during Operation Iraqi Freedom." Issue Paper. *Center for Strategic Leadership, U.S. Army War College* 8 (3). Retrieved from http://www.iwar.org.uk/news-archive/iraq/oif-reporters.pdf.

Perez-Peña, Richard. 2007. "News Coverage Shifts to Election from War in Iraq." *New York Times,* August 20.

Pietroluongo, Silvio, Minal Patel, and Wade Jessen. 2001. "Singles Minded." *Billboard,* November 17. Retrieved from LexisNexis Database.

———. 2002. Singles Minded. *Billboard,* August 3. Retrieved from LexisNexis Database.

———. 2003. Singles Minded. *Billboard,* May 10. Retrieved from LexisNexis Database.

Potter, Deborah. 2004. "Viewer Beware." *American Journalism Review* 26 (1): 60.

Project for Excellence in Journalism. 2003. "Jessica Lynch: Media Mythmaking in the Iraq War." *Journalism.org,* June 23. Retrieved from http://www.journalism.org/node/223.

———. 2004. *The State of the News Media 2004: An Annual Report on American Journalism.* Washington, D.C.: Project for Excellence in Journalism. Retrieved from http://www.stateofthenewsmedia.org/2004.

————. 2005. *The State of the News Media 2005: An Annual Report on American Journalism.* Washington, D.C.: Project for Excellence in Journalism. Retrieved from http://www.stateofthemedia.org/2005.

————. 2008. *The State of the News Media 2008: An Annual Report on American Journalism.* Washington, D.C.: Project for Excellence in Journalism. Retrieved from http://www.stateofthenewsmedia.org/2008.

Proffitt, Jennifer M. 2004. "Rupert Murdoch's Challenge to Democracy: Concentration and the Marginalization of Dissent." Unpublished Paper, College of Communications, Pennsylvania State University.

Propp, Vladimir. 1970. *Morphology of the Folktale.* Austin: University of Texas Press.

Purdum, Todd S., and Jim Rutenberg. 2003. "Reporters Respond Eagerly to Pentagon Welcome Mat." *New York Times,* March 23.

Rabinovitz, Lauren, and Susan Jeffords. 1994. "Introduction." In *Seeing through the Media: The Persian Gulf War,* ed. Susan Jeffords and Lauren Rabinovitz, 1–18. Rutgers, N.J.: Rutgers University Press.

Radyuhin, Vladimir. 2003. "Russia, France, Germany Stick to 'No-War' Stand." *The Hindu,* March 11. Retrieved from LexisNexis Database.

Rampton, Sheldon, and John Stauber. 2003. *Weapons of Mass Deception: The Uses of Propaganda in Bush's War on Iraq.* New York: Jeremy P. Tarcher/Penguin.

Rapping, Elayne. 1987. *The Looking Glass World of Nonfiction TV.* Boston, Mass.: South End Press.

Robinson, Piers. 2002. *The CNN Effect: The Myth of News, Foreign Policy and Intervention.* London and New York: Routledge.

Romano, Allison. 2003. "Ratings Alert: War Gives Fox News a Big Bump." *Broadcasting & Cable,* March 31. Retrieved from http://www.broadcastingcable.com/article/CA288044.html ?display=Top+of+the+Week.

Rosen, Jill. 2002. "Making Some Noise." *American Journalism Review* 24 (3): 12–13.

Rowse, Arthur E. 2000. *Drive-By Journalism: The Assault on Your Need to Know.* Monroe, Me.: Common Courage Press.

Ruoff, Jeffrey K. 1992. "Conventions of Sound in Documentary." In *Sound Theory, Sound Practice,* ed. Rick Altman, 217–234. New York: Routledge.

Rutenberg, Jim. 2003. "Cable's War Coverage Suggests a New 'Fox Effect' on Television Journalism." *New York Times,* April 16.

Rutherford, Paul. 2004. *Weapons of Mass Persuasion: Marketing the War against Iraq.* Toronto: University of Toronto Press.

Said, Carolyn. 2003. "Marketing Experts Say War Is a Tough Sell; Sound Bites, Slogans Strive for Image of Quick, Clean War." *San Francisco Chronicle,* March 30.

Said, Edward. 2003/1978. *Orientalism.* New preface by Edward Said. New York: Vintage Books.

————. 1997. *Covering Islam: How the Media and the Experts Determine How We See the Rest of the World.* Updated ed., with a new intro. New York: Vintage Books.

Schatz, Thomas. 1997. "The Return of the Hollywood Studio System." In Eric Barnouw et al., *Conglomerates and the Media,* 73–106. New York: New Press.

Schiller, Herbert I. 1992. *Mass Communications and American Empire.* 2nd ed. Boulder, Colo.: Westview Press.

Schudson, Michael. 1995. *The Power of News.* Cambridge, Mass.: Harvard University Press.

Seiter, Ellen. 1992. "Semiotics, Structuralism, and Television." In *Channels of Discourse, Reassembled,* ed. Robert C. Allen, 31–66. 2nd ed. Chapel Hill: University of North Carolina Press.

Sharkey, Jacqueline E. 2003. "The Rise of Arab TV." *American Journalism Review* 25 (4): 26–27.

———. 2004. "Al-Jazeera under the Gun." *American Journalism Review* 26 (5): 18–19.

Shohat, Ella. 1994. "The Media's War." In *Seeing through the Media: The Persian Gulf War,* ed. Susan Jeffords and Lauren Rabinovitz, 147–154. Rutgers, N.J.: Rutgers University Press.

Shohat, Ella, and Robert Stam. 1994. *Unthinking Eurocentrism: Multiculturalism and the Media.* New York: Routledge.

Sperry, Sharon Lynn. 1981. "Television News as Narrative." In *Understanding Television: Essays on Television as a Social and Cultural Force,* ed. Richard P. Adler, 295–312. New York: Praeger.

Stam, Robert. 1983. "Television News and Its Spectator." In *Regarding Television—Critical Approaches: An Anthology,* ed. E. Ann Kaplan, 23–43. Frederick, Md.: University Publications of America.

Steinberg, Jacques. 2008. "Fox News Finds Its Rivals Closing In." *New York Times,* June 28.

Sterngold, James. 1998. "Journalism Goes Hollywood, and Hollywood Is Reading." *New York Times,* July 10.

Strauss, Gary. 2008. "'Kill' In Search of a Target." *USA Today,* July 10. Retrieved from http://findarticles.com/p/articles/mi_kmusa/is_/ai_n27916232.

Taranto, James. 2005. "Toy Soldiers." *Wall Street Journal,* February 2. Retrieved from http://www.opinionjournal.com/best/?id=110006241.

Taylor, Philip M. 1998. *War and the Media.* Manchester: Manchester University Press.

Thielman, Sam. 2008. "Olbermann Factors In." *Daily Variety,* June 11. Retrieved from LexisNexis Database.

Thompson, Kristin, and David Bordwell. 1994. *Film History: An Introduction.* New York: McGraw-Hill.

Thussu, Daya Kishan. 2003. "Live TV and Bloodless Deaths: War, Infotainment, and 24/7 News." In *War and the Media: Reporting Conflict 24/7,* ed. Daya Kishan Thussu and Des Freedman, 117–132. Thousand Oaks, Calif.: Sage Publications.

TimeWarner.com. 2005. "Newsroom: CNN/US Showcases Global Resources with *Your World Today* Simulcast." May 10. Retrieved from http://www.timewarner.com/corp/newsroom/pr/0,20812,1059861,00.html.

Tomkins, Richard. 2003. "No Logo Can Sell the True Horrors of Battle." *Financial Times,* March 20.

Tracy, Paula, and Katharine McQuaid. 2003. "NH Man's Flag Began Journey on 9/11." *Manchester Union Leader* (Manchester, N.H.), April 12. Retrieved from LexisNexis Database.

U.S. Department of Defense. 2003a. News Transcript: DoD News Briefing—Secretary Rumsfeld and Gen. Myers. March 11. Retrieved from http://www.defenselink.mil/transcripts/2003/t03112003_t0311sd.html.

————. 2003b. News Transcript: Deputy Secretary Wolfowitz Interview with *Newsweek*. March 12. Retrieved from http://www.defenselink.mil/transcripts/transcript.aspx?transcriptid=2061.

————. 2003c. News Transcript: DoD News Briefing—Secretary Rumsfeld and Gen. Myers. March 20. Retrieved from http://www.defenselink.mil/transcripts/2003/t03202003_t0320sd.html.

————. 2003d. News Transcript: DoD News Briefing—Secretary Rumsfeld and Gen. Myers. March 21. Retrieved from http://www.defenselink.mil/transcripts/2003/t03212003_t0321sd1.html.

————. 2003e. News Transcript: DoD News Briefing—ASD PA Clarke and Maj. Gen. McChrystal. March 22. Retrieved from http://www.defenselink.mil/transcripts/2003/t03222003_t03220sdpa.html.

————. 2003f. News Transcript: DoD News Briefing—ASD PA Clarke and Maj. Gen. McChrystal. March 24. Retrieved from http://www.defenselink.mil/transcripts/transcript.aspx?transcriptid=2134.

————. 2003g. News Transcript: DoD News Briefing—Secretary Rumsfeld and Gen. Myers. April 9. Retrieved from http://www.defenselink.mil/transcripts/2003/tr20030409-secdef0084.html.

Virilio, Paul. 1984. *War and Cinema: The Logistics of Perception.* Trans. Patrick Camiller. London: Verso.

————. 2002. *Desert Screen: War at the Speed of Light.* Trans. Michael Degener. London: Continuum.

Von Drehle, David, and R. Jeffrey Smith. 1993. "U.S. Strikes Iraq for Plot to Kill Bush." *Washington Post,* June 27.

Von Rhein, John. 2003. "Networks' Theme Music Sanitizes War's Darkest Realities." April 6. Retrieved from http://www.worldofradio.com/dxld3061.txt.

Wasko, Janet. 1994. *Hollywood in the Information Age.* Austin: University of Texas Press.

White, Mimi. 1992. "Ideological Analysis and Television." In *Channels of Discourse, Reassembled,* ed. Robert C. Allen, 161–202. 2nd ed. Chapel Hill: University of North Carolina Press.

White House. 2003a. Press Briefing by Ari Fleischer. March 20. http://www.whitehouse.gov/news/releases/2003/03/20030320-3.html#15 (accessed November 15, 2005).

————. 2003b. Press Briefing by Ari Fleischer. March 21. http://www.whitehouse.gov/news/releases/2003/03/20030321-9.html#3 (accessed November 15, 2005).

————. 2003c. Press Briefing by Ari Fleischer. March 24. http://www.whitehouse.gov/news/releases/2003/03/20030324-4.html#2b (accessed November 15, 2005).

————. 2003d. Press Release: Coalition Members. March 25. http://www.whitehouse.gov/news/releases/2003/03/20030325-9.html (accessed November 15, 2005).

Williams, Raymond. 1974. *Television: Technology and Cultural Form.* New York: Schocken Books; repr., London: Routledge Classics, 2005.

Wittebols, James H. 2004. *The Soap Opera Paradigm: Television Programming and Corporate Priorities.* Lanham, Md.: Rowman & Littlefield Publishers.

Wolff, Michael. 2003. "My Big Fat Question." *New York Magazine,* April 21. Retrieved from http://www.newyorkmetro.com/nymetro/news/media/columns/medialife/n_8623/index.html.

Wood, Robin. 1986. *Hollywood from Vietnam to Reagan.* New York: Columbia University Press.

———. 1995. "Ideology, Genre, Auteur." In *Film Genre Reader II,* ed. Barry Keith Grant, 59–73. Austin: University of Texas Press.

Woodward, Gary C. 1997. *Perspectives on American Political Media.* Boston: Allyn & Bacon.

WorldWatch Institute. 2003. "Between the Lines: What the Iraq War Costs Could Have Bought." *World Watch Magazine,* May/June.

Wright, Judith Hess. 1995. "Genre Films and the Status Quo." In *Film Genre Reader II,* ed. Barry Keith Grant, 41–49. Austin: University of Texas Press.

Wu, Steven. 1999. "This Just In: Qatar's Satellite Channel." *Harvard International Review* 21 (4): 14–15.

Wurtzler, Steve. 1992. "'She Sang Live, but the Microphone Was Turned Off': The Live, the Recorded and the *Subject* of Representation." In *Sound Theory, Sound Practice,* ed. Rick Altman, 87–103. New York: Routledge.

Wyatt, Justin. 1994. *High Concept: Movies and Marketing in Hollywood.* Austin: University of Texas Press.

Xinhua General News Service. 2003. "Chirac, Putin Reaffirm Convergence of Position on Iraq." March 4. Retrieved from the Lexis/Nexis Database.

Ytreberg, Espen. 2001. "Moving out of the Inverted Pyramid: Narratives and Descriptions in Television News." *Journalism Studies* 2 (3): 357–371.

Zednik, Rick. 2002. "Inside Al Jazeera." *Columbia Journalism Review* 40, no. 6.

### Sound Recordings

Black, Clint. 2003. "Iraq and I Roll." Available online only. Retrieved from http://www.pro-american.com/Free_Music/free_music.html (accessed September 2005).

Charlie Daniels Band. 2001. "This Ain't No Rag, It's a Flag." From *Patriotic Country.* Compact Disc. BMG Heritage.

Keith, Toby. 2002. "Courtesy of the Red, White, and Blue (The Angry American)." From *Unleashed.* Compact Disc. Dreamworks Nashville.

Worley, Darryl. 2003. "Have You Forgotten?" From *Have You Forgotten?* Compact Disc. Dreamworks Nashville.

# Index

**Deborah L. Jaramillo** is Assistant Professor in the Department of Film and Television at Boston University.